D1078036

end

Edexcel

Biology for **AS**

C J Clegg

NORWICH CITY COLLEGE		
Stock No.	244 294	
Class	570	CLE
Cat.	BZ	Proc	3wr

DYNAMIC
LEARNING

244 294

HODDER
EDUCATION
AN HACHETTE UK.COMPANY

Although every effort has been made to ensure that website addresses are correct at time of going to press, Hodder Education cannot be held responsible for the content of any website mentioned in this book. It is sometimes possible to find a relocated web page by typing in the address of the home page for a website in the URL window of your browser.

Hachette UK's policy is to use papers that are natural, renewable and recyclable products and made from wood grown in sustainable forests. The logging and manufacturing processes are expected to conform to the environmental regulations of the country of origin.

Orders: please contact Bookpoint Ltd, 130 Milton Park, Abingdon, Oxon OX14 4SB. Telephone: (44) 01235 827720. Fax: (44) 01235 400454. Lines are open 9.00–5.00, Monday to Saturday, with a 24-hour message answering service. Visit our website at www.hoddereducation.co.uk

© C.J. Clegg 2008
First published in 2008 by
Hodder Education,
An Hachette UK Company
338 Euston Road
London NW1 3BH

Impression number   6
Year   2012

All rights reserved. Apart from any use permitted under UK copyright law, no part of this publication may be reproduced or transmitted in any form or by any means, electronic or mechanical, including photocopying and recording, or held within any information storage and retrieval system, without permission in writing from the publisher or under licence from the Copyright Licensing Agency Limited. Further details of such licences (for reprographic reproduction) may be obtained from the Copyright Licensing Agency Limited, Saffron House, 6–10 Kirby Street, London EC1N 8TS.

Cover photo: dividing *Streptococcus* bacteria, CNRI/Science Photo Library
Illustrations by Oxford Designers and Illustrators
Typeset in 10pt Goudy by Fakenham Prepress Solutions, Fakenham, Norfolk NR21 8NN
Printed in Dubai

A catalogue record for this title is available from the British Library

ISBN: 978 0 340 96623 5

Author's acknowledgements

I am indebted to Dr Neil Millar of Heckmondwike Grammar School, Kirklees, West Yorkshire, for creating the Appendix *Handling Data* to meet the specific requirements of AS Biology students.

I have had the benefit of advice and insights on the specific requirements of the new Edexcel Specification from Mr Ed Lees, previously Head of Sixth Form at The Ridgeway School, Wroughton, and Principal Examiner and Principal Moderator for Edexcel. As an experienced teacher of biology, his perceptive observations on content and approach of the manuscript as a whole were invaluable. Nevertheless, any remaining inaccuracies are my sole responsibility. I hope readers will write to point out any faults they find.

At Hodder Education, the skill and patience of Katie Mackenzie Stuart (Science Publisher), Helen Townson (Designer), Rebecca Teevan (Picture Research), and my Freelance Editor Gina Walker have combined to bring together materials as I have wished, and I am most grateful to them.

Dr Chris Clegg,
Salisbury, Wiltshire, May, 2008.

Practical work

In this text and the accompanying Dynamic Learning resources, there are suggestions for practical work which teachers might like students to perform. Detailed procedures can be found in other texts.

Since 1989, risk assessment has been required by law, initially by the Control of Substances Hazardous to Health (COSHH) Regulations but now by several sets of legislation, including the Management of Health and the Personal Protective Equipment (PPE) Regulations. All of these come under the umbrella of the Health and Safety at Work etc. Act. However, from a practical point of view, it matters little under which regulations a risk assessment is to be carried out, so that the requirement can be summed up as follows:

> A risk assessment is needed for any activity in which there is a significant hazard, whether carried out by pupils, teachers or technicians.

For those schools that subscribe to CLEAPSS (www.cleapss.org.uk) either through their Local Authority or as Independant members, they should consult CLEAPSS publications or contact CLEAPSS directly on the *Helpline*. It is advisable that a science safety policy for each establishment is written for the benefit of teachers and technicians (see Guide L196 and guide L223).

A risk assessment must be made for any work where students are the subject of an investigation (e.g. recording heartbeats), where microorganisms are involved (see section 15 of the Handbook) or where chemicals are handled (use Hazcards).

Using the Dynamic Learning resources

To access **Activities**, **How Science Works** resources and **A* Extension** materials, move the cursor over the yellow box in the margin, and left click. The text appears in a pop-up window – click on the PDF icon in the top left corner to open the resource.

Click on a **Figure** to enlarge it in a pop-up window. Then you can zoom into any part, and drag around the piece using the cursor. You can save **illustrations** or **photographs** to your own folders, by clicking on the icons in the top left corner.

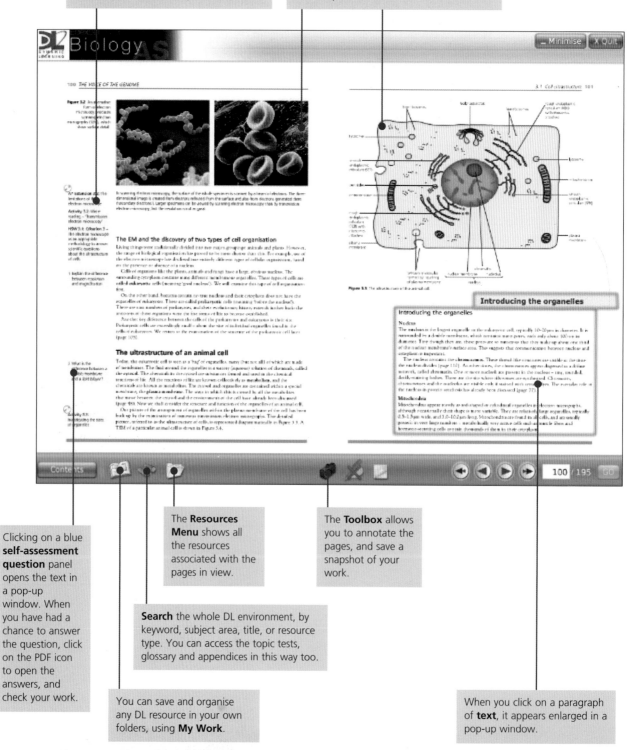

Clicking on a blue **self-assessment question** panel opens the text in a pop-up window. When you have had a chance to answer the question, click on the PDF icon to open the answers, and check your work.

The **Resources Menu** shows all the resources associated with the pages in view.

The **Toolbox** allows you to annotate the pages, and save a snapshot of your work.

Search the whole DL environment, by keyword, subject area, title, or resource type. You can access the topic tests, glossary and appendices in this way too.

You can save and organise any DL resource in your own folders, using **My Work**.

When you click on a paragraph of **text**, it appears enlarged in a pop-up window.

Contents

Full **End-of-topic tests** and a comprehensive **Glossary** are available via the DL Student website, along with **Activities, How science works** (HSW) inputs, **A* Extensions, Answers** to SAQs and two **Appendices: Background Chemistry for Biologists**, and **Handling Data.**
(To access these resources in Dynamic Learning, use the 'search' tool, whose icon appears at the left end of the task bar. Type your requirements in the box and click 'go'. The resources are also accessible via pop-up boxes on relevant pages.)

Introduction

Edexcel Biology for AS is designed to deliver the learning outcomes of the Concept approach of the Edexcel Specification. The intentions are to:

- develop knowledge of biology, together with the skills needed to use it
- facilitate an understanding of scientific methods
- sustain and develop a life-long interest and enjoyment in biology
- recognise the value of biology in society, so that it may be used responsibly.

The special features of this presentation include:

- explanations commence at the level required for a grade C at GCSE Science, and chapters begin with **Starting points** which identify these basic issues
- the text responds to the given learning objectives of the topics, more-or-less in sequence, and is written in straightforward language. Biology has a demanding vocabulary, so essential terms are explained as they arise, and reminders are given in the **Glossary**
- photographs, electron micrographs and full-colour illustrations are linked to support the text, with annotations included to elaborate the context, function or applications
- self-assessment questions support comprehension or require research of interconnecting ideas, throughout the chapters. **End-of-topic tests** are given to support revision. The Edexcel website (www.edexcel.com/gce2008) provides sample assessment materials to consult (with mark schemes available).
- throughout the text, additional resources are flagged, and accessed via the interactive pages of the **DL Student website** DL, including:

1 **Activities**, consisting of demonstrations, ideas for investigations, further reading sources, website addresses, and pencil-and-paper tasks that challenge understanding
2 **How science works** (HSW) inputs that focus on specific criteria illustrated by issues in the text. Of course, HSW issues arise widely, especially within learning objectives involving practical work
3 **A* Extensions**, which take opportunities to discuss wider or more challenging applications, knowledge of which is not demanded but provides a framework for greater involvement.

Also available via the DL Student website DL are the **Glossary**, **Answers** to self-assessment questions in the text, and two **Appendices**:

- *Background Chemistry for Biologists*
- *Handling Data*, written by Guest Author Dr Neil Millar. Here is attached his unique statistical package *Merlin*, created for biology students, which is made freely available for educational and non-profit use.

1 Lifestyle, health and risk

STARTING POINTS
- Water, a major component of living things, has a unique molecular structure. Many of the properties of water are essential for life.
- Compounds built from carbon are called organic compounds, of which vast numbers make up living things. Most fall into one of four groups of compounds – these are carbohydrates, lipids (for example, fats and oils), proteins and nucleic acids.
- Cells that make up organisms require a supply of nutrients and water, and most require oxygen. Waste products must be disposed of. Across large, multicellular organisms, movement of these substances by diffusion alone is not sufficient – an internal transport system is required.
- Good health is more than the absence of disease; ill-health may be caused by unfavourable environmental conditions, too. Unhealthy lifestyles increase the risks of disease and body malfunction.

1.1 Introducing biological molecules

Chemical **elements** are the units of pure substance that make up our world. The Earth is composed of about 92 stable elements in all, in varying quantities, and living things are built from some of them. Table 1.1 shows a comparison of the most common elements in the Earth's crust and in us. You can see that the bulk of the Earth is composed of the elements oxygen, silicon, aluminium and iron. Of these, only oxygen is a major component of our cells.

Table 1.1 The most common elements in the Earth's crust and in the human body, in descending order of abundance (as percentages of the total number of atoms).

Earth's crust		Human body	
oxygen	47%	hydrogen	63%
silicon	28%	oxygen	25.5%
aluminium	7.9%	carbon	9.5%
iron	4.5%	nitrogen	1.4%
calcium	3.5%	calcium	0.31%
sodium	2.5%	phosphorus	0.22%

In fact, about 16 elements are required to build up all the molecules of the cell, and are therefore essential for life. Consequently, the full list of essential elements is a relatively short one. Furthermore, about 99% of living matter consists of just four elements: carbon, hydrogen, oxygen and nitrogen.

Why do these four elements predominate in living things?

The elements carbon, hydrogen and oxygen predominate because living things contain large quantities of water, and also because most other molecules present in cells and organisms are compounds of carbon combined with hydrogen and oxygen, including the **carbohydrates** and **lipids**. The element nitrogen is combined with carbon, hydrogen and oxygen in compounds called **amino acids** from which **proteins** are constructed. We will discuss water, carbohydrates and lipids first.

Water

Living things are typically solid, substantial objects, yet water forms the bulk of their structures – between 65% and 95% by mass of most multicellular plants and animals (about 80% of a human cell consists of water). Despite this, and the fact that water has some unusual properties, it is a substance that is often taken for granted.

Water is composed of atoms of the elements hydrogen and oxygen. One atom of oxygen and two atoms of hydrogen combine by sharing of electrons in an arrangement known as **covalent bonding** (Figure 1.3). However, the water molecule is triangular rather than linear, and the nucleus of the oxygen atom draws electrons (negatively charged) away from the hydrogen nuclei (positively charged) – with an interesting consequence. Although overall the water molecule is electrically neutral, there is a net **negative charge** on the oxygen atom and a net **positive charge** on the hydrogen atoms. In other words, the water molecule carries an unequal distribution of electrical charge within it. This arrangement is known as a **polar molecule** (Figure 1.1).

Figure 1.1 The water molecule and the hydrogen bonds it forms.

1 Explain the differences between the terms *atom* and *ion*.

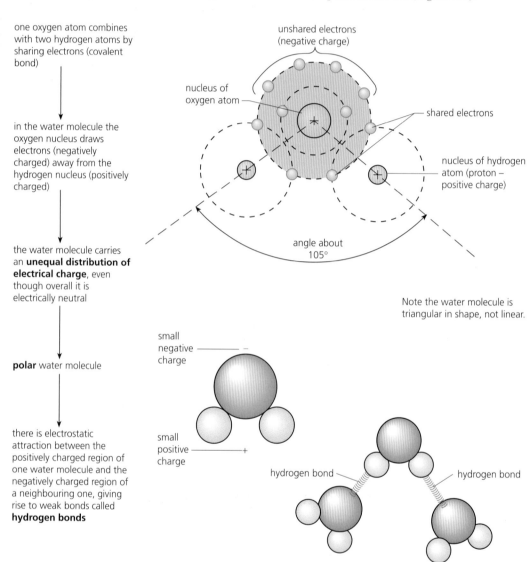

one oxygen atom combines with two hydrogen atoms by sharing electrons (covalent bond)

in the water molecule the oxygen nucleus draws electrons (negatively charged) away from the hydrogen nucleus (positively charged)

the water molecule carries an **unequal distribution of electrical charge**, even though overall it is electrically neutral

polar water molecule

there is electrostatic attraction between the positively charged region of one water molecule and the negatively charged region of a neighbouring one, giving rise to weak bonds called **hydrogen bonds**

unshared electrons (negative charge)

nucleus of oxygen atom

shared electrons

nucleus of hydrogen atom (proton – positive charge)

angle about 105°

Note the water molecule is triangular in shape, not linear.

small negative charge

small positive charge

hydrogen bond

hydrogen bond

Hydrogen bonds

With water molecules, the positively charged hydrogen atoms of one molecule are attracted to negatively charged oxygen atoms of nearby water molecules by forces called **hydrogen bonds**. These are weak bonds compared to covalent bonds, yet they are strong enough to hold water molecules together and attract water molecules to charged particles or charged surfaces. In fact, hydrogen bonds largely account for the unique properties of water, one of which is as a solvent, which is examined next.

Solvent properties of water

Water is a powerful solvent for polar substances. These include:

- ionic substances like sodium chloride (Na^+ and Cl^-) – all **cations** (positively charged ions) and **anions** (negatively charged ions) become surrounded by a shell of orientated water molecules (Figure 1.2)
- carbon-containing (organic) molecules with ionised groups (such as the carboxyl group $-COO^-$, and amino group $-NH^{3+}$) – soluble organic molecules like sugars dissolve in water due to the formation of hydrogen bonds with their slightly charged hydroxyl groups ($-OH$).

Once they have dissolved, molecules of the substance (**solute**) are free to move around in the water (**solvent**), and as a result, are more chemically reactive than when in the undissolved solid.

On the other hand, non-polar substances are repelled by water, as in the case of oil on the surface of water. Non-polar substances are **hydrophobic** (water-hating).

Figure 1.2 Water as universal solvent.

Ionic compounds like NaCl dissolve in water,

$$NaCl \rightleftharpoons Na^+ + Cl^-$$

with a group of orientated water molecules around each ion:

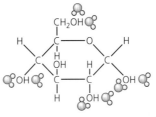

2 In an aqueous solution of glucose, which component is the solvent and which is the solute?

3 Why is a hydrogen ion also referred to as a proton?

Sugars and alcohols dissolve due to hydrogen bonding between polar groups in their molecules (e.g. —OH) and the polar water molecules:

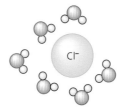

Activity 1.1: Water as a polar molecule – a demonstration

HSW 1.1: Criterion 5 – Water as a liquid at room temperature

A* Extension 1.1: Water, a unique molecule with properties that 'life' depends on

4 Suggest where the non-organic forms of carbon most commonly occur.

The carbon of organic compounds

Carbon is a relatively uncommon element of the Earth's crust, but in cells and organisms it is the second most abundant element by mass, after oxygen. The majority of the carbon compounds found in living organisms are relatively large molecules in which many carbon atoms are combined together, with hydrogen and oxygen. They are known as **organic compounds**. The exceptions are the gas carbon dioxide (CO_2), hydrogen carbonate ions (the product of CO_2 when dissolved in water), and mineral calcium carbonate ($CaCO_3$), which are classified as 'non-organic' forms of carbon.

Carbon has remarkable properties. It is a relatively small atom, but it is able to form four strong, stable, covalent bonds (Figure 1.3). These bonds point to the corners of a regular tetrahedron (a pyramid with a triangular base). This is because the four pairs of electrons repel each other and so position themselves as far away from each other as possible.

Carbon atoms are able to react with each other to form extended chains. The resulting carbon 'skeletons' may be straight chains, branched chains, or rings. Carbon also bonds covalently with other atoms, such as oxygen, hydrogen, nitrogen and sulphur, forming different groups of organic molecules with distinctive properties. These features of carbon are introduced in Figure 1.3.

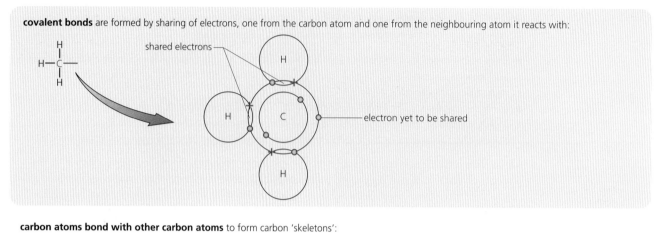

covalent bonds are formed by sharing of electrons, one from the carbon atom and one from the neighbouring atom it reacts with:

shared electrons

electron yet to be shared

carbon atoms bond with other carbon atoms to form carbon 'skeletons':

straight

short   long   branched   or ring forms

short chain (in the
amino acid alanine)

long chain (in a fatty acid)

branched chain (in the
amino acid valine)

the ring form
(of α-glucose)

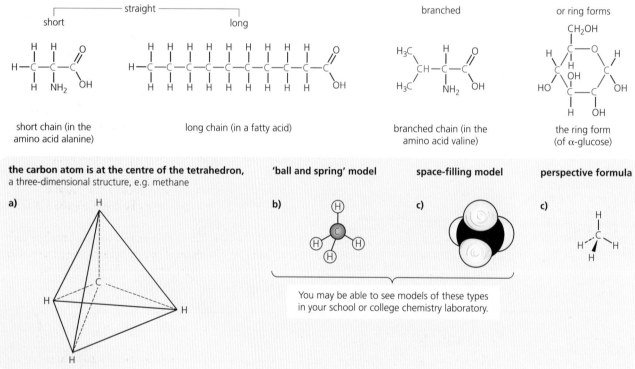

the carbon atom is at the centre of the tetrahedron, a three-dimensional structure, e.g. methane

a)

'ball and spring' model

b)

space-filling model

c)

perspective formula

c)

You may be able to see models of these types
in your school or college chemistry laboratory.

Figure 1.3 The tetrahedral carbon atom, its covalent bonds and the carbon 'skeletons' it may form.

One inevitable outcome of these features is that there are vast numbers of organic compounds – more than the total of known compounds made from other elements, in fact. Biologists think the diversity of organic compounds has made possible the diversity of life. Fortunately, very many of the organic chemicals of living things fall into one of four discrete groups or 'families' of chemicals with many common properties, one of which is the **carbohydrates**. We will consider this family of molecules first.

Carbohydrates

Carbohydrates are the largest group of organic compounds found in living things. They include sugars, starch, glycogen and cellulose. Carbohydrates contain only three elements: carbon, hydrogen and oxygen, with hydrogen and oxygen always present in the ratio $2:1$ (as they are in water, H_2O), so they can be represented by the general formula $C_x(H_2O)_y$. Table 1.2 is a summary of the three types of carbohydrates of living things.

Table 1.2 Carbohydrates of cells and organisms, general formula $C_x(H_2O)_y$.

	Monosaccharides	Disaccharides	Polysaccharides
Description	simple sugars	two simple sugars chemically combined together	very many simple sugars chemically combined together
Examples	glucose (six carbon atoms) fructose (six carbon atoms) ribose (five carbon atoms)	sucrose lactose maltose	starch glycogen cellulose

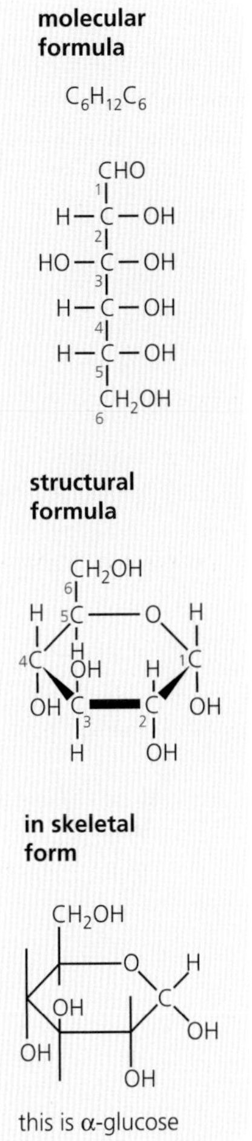

molecular formula

$C_6H_{12}C_6$

structural formula

in skeletal form

this is α-glucose

Figure 1.4 The structure of glucose.

Monosaccharides – the simple sugars

Monosaccharides are carbohydrates with relatively small molecules. They are soluble in water and taste sweet. In biology, **glucose** is an especially important monosaccharide because:

- all green leaves manufacture glucose using light energy
- our bodies transport glucose in the blood
- all cells use glucose in respiration – it is one of the respiratory substrates
- in cells and organisms, glucose is the building block for very many larger molecules.

The structure of glucose

Glucose has a chemical or **molecular formula** of $C_6H_{12}O_6$. This type of formula tells us what the component atoms are, and the numbers of each atom in the molecule. For example, glucose is a six-carbon sugar, or **hexose**. But the molecular formula does not tell us the structure of the molecule.

Glucose can be written on paper as a linear molecule but it cannot exist in this form. *Can you see why?*

It's because each carbon arranges its four bonds into a tetrahedron (Figure 1.3), so the molecule cannot be 'flat'. Rather, glucose is folded, taking a ring or **cyclic form**. Figure 1.4 shows the **structural formula** of glucose.

The carbon atoms of an organic molecule may be numbered. This allows us to identify which atoms are affected when the molecule reacts and changes shape. For example, as the glucose ring forms, the oxygen on carbon-5 attaches itself to carbon-1. Note that the glucose ring contains five carbon atoms and an oxygen atom (Figure 1.4).

■ Extension: Isomers of glucose

Compounds that have the same chemical formula (the same component atoms in their molecules) but that differ in the arrangement of the atoms are known as **isomers**. Many organic compounds exist in isomeric forms, and so it is often important to know the structure of an organic compound as well as its composition. (The names of organic compounds, and of the linkages between compounds when they combine, are sometimes complex because they give this information.)

In the ring structure of glucose, the positions of –H and –OH that are attached to carbon-1 may interchange, giving rise to two isomers, known as α-glucose and β-glucose. When we compare the structure of the polysaccharide cellulose (Chapter 4, page 145–6) with that of starch and glycogen (below), the significance of this difference will become apparent. In the meantime, we use these terms in describing the linkages in starch (amylose and amylopectin) and glycogen. Now you know what these names mean.

5 Distinguish between *ionic* and *covalent bonding*.

Other monosaccharides of importance in living cells

Glucose, fructose and galactose are examples of hexose sugars commonly occurring in cells and organisms. Other monosaccharide sugars produced by cells and used in metabolism include five-carbon sugars (**pentoses**), namely ribose and deoxyribose. These are components of the nucleic acids (Chapter 2).

Disaccharides

Disaccharides are carbohydrates made of two monosaccharides combined together. For example, **sucrose** is formed from a molecule of glucose and a molecule of fructose chemically combined together.

What type of chemical reaction is involved?

Condensation and hydrolysis reactions

When two monosaccharide molecules are combined to form a disaccharide, a molecule of water is also formed as a product, and so this type of reaction is known as a **condensation reaction**. The linkage between monosaccharide residues, after the removal of H–O–H between them, is called a **glycosidic linkage** (Figure 1.5). This is comprised of strong, covalent bonds. The condensation reaction is brought about by an enzyme (Chapter 2).

In the reverse process, disaccharides are 'digested' to their component monosaccharides in a **hydrolysis reaction**. Of course, this reaction involves adding a molecule of water (*hydro-*), as splitting (*-lysis*) of the glycosidic linkage occurs. It is catalysed by an enzyme too, but it is a different enzyme from the one that brings about the condensation reaction.

Apart from sucrose, other disaccharide sugars produced by cells and used in metabolism include:

- **maltose**, formed by condensation reaction of two molecules of glucose
- **lactose**, formed by condensation reaction of galactose and glucose.

Figure 1.5 Disaccharides, and the monosaccharides that form them.

6 Suggest why simple sugars like glucose are not commonly found as a storage form of carbohydrate in cells or tissues.

sucrose + water $\underset{\text{condensation}}{\overset{\text{hydrolysis}}{\rightleftharpoons}}$ glucose + fructose

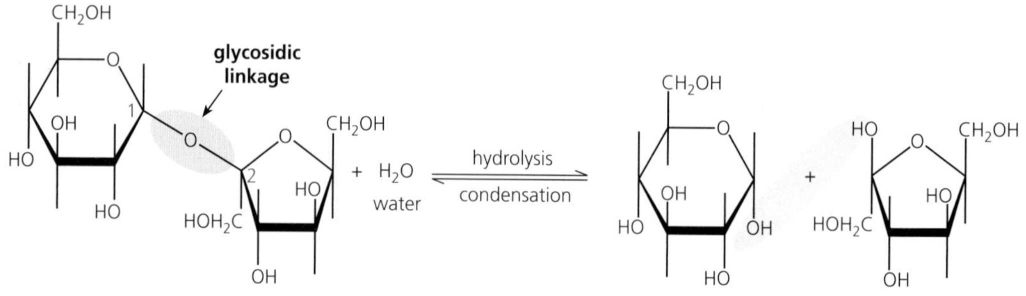

This structural formula shows us how the glycosidic linkage forms/breaks.

maltose + water $\underset{\text{condensation}}{\overset{\text{hydrolysis}}{\rightleftharpoons}}$ glucose + glucose

lactose + water $\underset{\text{condensation}}{\overset{\text{hydrolysis}}{\rightleftharpoons}}$ galactose + glucose

Polysaccharides

A **polysaccharide** is built from many monosaccharides connected by glycosidic linkages formed in condensation reactions. *Poly* means 'many', and in fact thousands of saccharide (sugar) units make up a polysaccharide. So a polysaccharide is a giant molecule, a **macromolecule**. Normally, each polysaccharide contains only one type of **monomer**. A chemist calls this a **polymer** because it is constructed from a huge number of *identical* monomers.

Some polysaccharides function as stores of energy; glycogen and starch are examples, as we shall shortly see. Other polysaccharides, such as chitin and cellulose, have a structural role. These are huge molecules that are not so easily hydrolysed by enzyme action.

Starch

Starch is a mixture of two polysaccharides, both of which are polymers of α-glucose. The **amylose** molecule is an unbranched chain, whereas **amylopectin** has branches at points along its chain. The linkages between glucose residues in starch bring the sugar molecules together so a helix forms (Figure 1.6). The whole starch molecule is then stabilised by countless hydrogen bonds between parts of the component glucose molecules.

Starch is the major storage carbohydrate of most plants. It is laid down as compact grains. Starch is an important energy source in the diet of many animals, too. It is useful because its molecules are both compact and insoluble, but are readily hydrolysed to form sugar when required. Of course, enzymes are concerned in this reaction, too.

We sometimes see 'soluble starch' as an ingredient of manufactured foods. Here the starch molecules have been broken down into short lengths, making them more easily dissolved.

We test for starch by adding a solution of iodine in potassium iodide. Iodine molecules fit neatly into the centre of the starch helix, creating a blue–black colour (Figure 1.6).

Figure 1.6 Starch

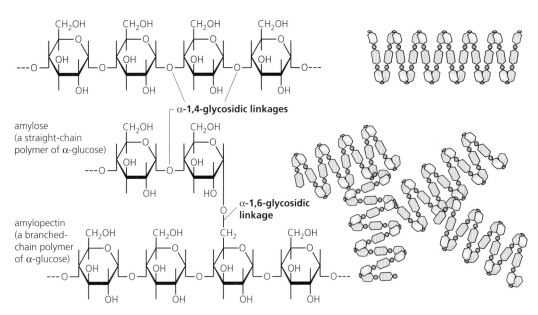

amylose (a straight-chain polymer of α-glucose)

α-1,4-glycosidic linkages

amylopectin (a branched-chain polymer of α-glucose)

α-1,6-glycosidic linkage

test for starch with iodine in potassium iodide solution, the blue-black colour comes from a starch/iodine complex

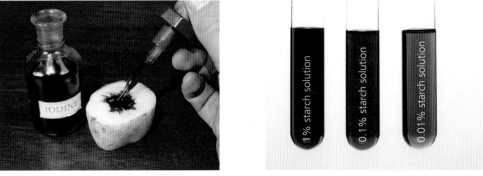

a) on a potato tuber cut surface

1% starch solution

0.1% starch solution

0.01% starch solution

b) on starch solutions of a range of concentrations

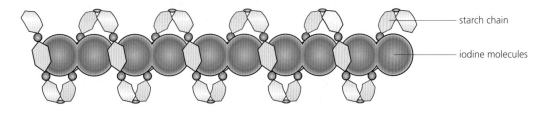

starch chain

iodine molecules

Glycogen

Glycogen is a polymer of α-glucose, chemically very similar to amylopectin, although larger and more highly branched (Figure 1.7). Granules of glycogen are seen in liver cells and muscle fibres, when observed using the electron microscope, but they occur throughout the human body, except in the brain cells (where there are virtually no energy reserves). During prolonged and vigorous exercise we draw on our glycogen reserves first. Only when these are exhausted does the body start to metabolise stored fat.

Figure 1.7 In glycogen, most glucose residues are linked by α-1,4 glycosidic linkages, creating a helical structure similar to amylopectin, but with more frequent branch points (α-1,6 glycosidic linkages).

Structural formula

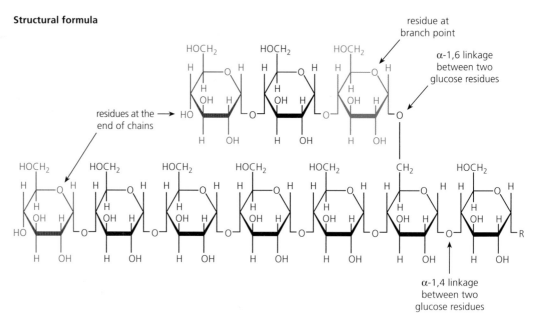

TEM of a liver cell (×7000)

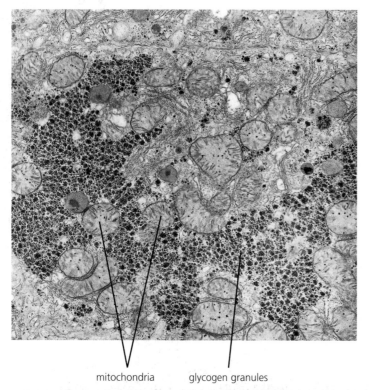

mitochondria   glycogen granules

Diagram to show the branching pattern of a glycogen molecule

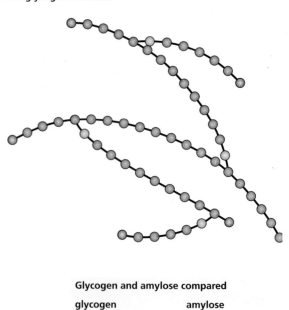

Glycogen and amylose compared

glycogen	amylose
branch point every 10 glucose residues	branch point every 30 glucose residues

Activity 1.2: Wider reading – 'Glucose and glucose-containing carbohydrates'

Activity 1.3: Molecular structure simulations

7 Define the terms *monomer* and *polymer*, giving examples from the carbohydrates.

Lipids

The second group of organic molecules to consider here are the **lipids**. These occur in living things as animal **fats** and plant **oils**. Fats and oils seem rather different substances, but their only difference is that at about 20 °C (room temperature) oils are liquid and fats are solid. Like the carbohydrates, lipids also contain the elements carbon, hydrogen and oxygen, but in lipids the proportion of oxygen is much smaller.

Lipids are insoluble in water. In fact, they generally behave as 'water-hating' molecules, and are described as **hydrophobic**. However, lipids can be dissolved in organic solvents such as alcohol (for example, ethanol) and propanone (acetone).

Triglycerides

Fats and oils are compounds called **triglycerides**. They are formed by reactions in which water is removed (another case of a condensation reaction) between fatty acids and an alcohol called glycerol. Here the link formed is known as an **ester linkage**.

The fatty acids combined in fats and oils have long hydrocarbon 'tails', typically about 16–18 carbon atoms long, though they may contain any number from 14 to 22 carbon atoms.

The structures of a fatty acid commonly found in cells and of glycerol are shown in Figure 1.8, and the steps to triglyceride formation are outlined in Figure 1.9. Of course, enzymes bring about condensation reactions by which triglycerides are formed.

The hydrophobic properties of triglycerides are due to the hydrocarbon tails of the component fatty acids. A molecule of triglyceride is quite large, but relatively small when compared to polymer macromolecules such as starch. However, because of their hydrophobic properties, triglyceride molecules clump together (aggregate) into huge globules in the presence of water, making them appear to be macromolecules.

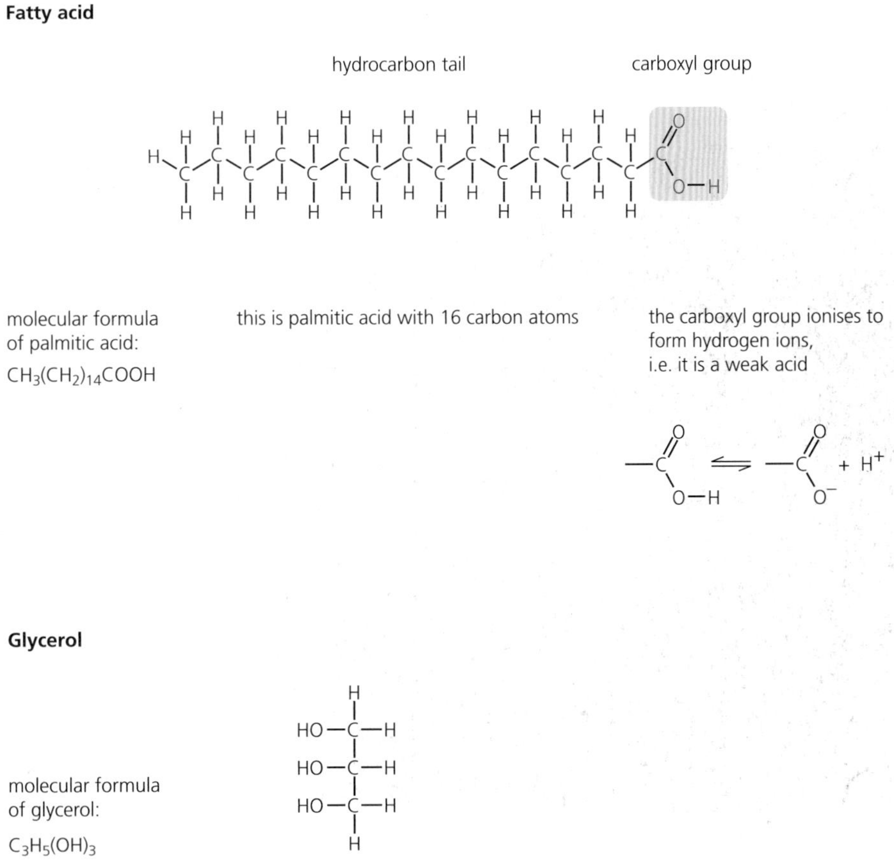

Fatty acid

hydrocarbon tail   carboxyl group

molecular formula of palmitic acid:

$CH_3(CH_2)_{14}COOH$

this is palmitic acid with 16 carbon atoms

the carboxyl group ionises to form hydrogen ions, i.e. it is a weak acid

Glycerol

molecular formula of glycerol:

$C_3H_5(OH)_3$

Figure 1.8 Fatty acids and glycerol, the building blocks.

Figure 1.9 Formation of triglyceride.

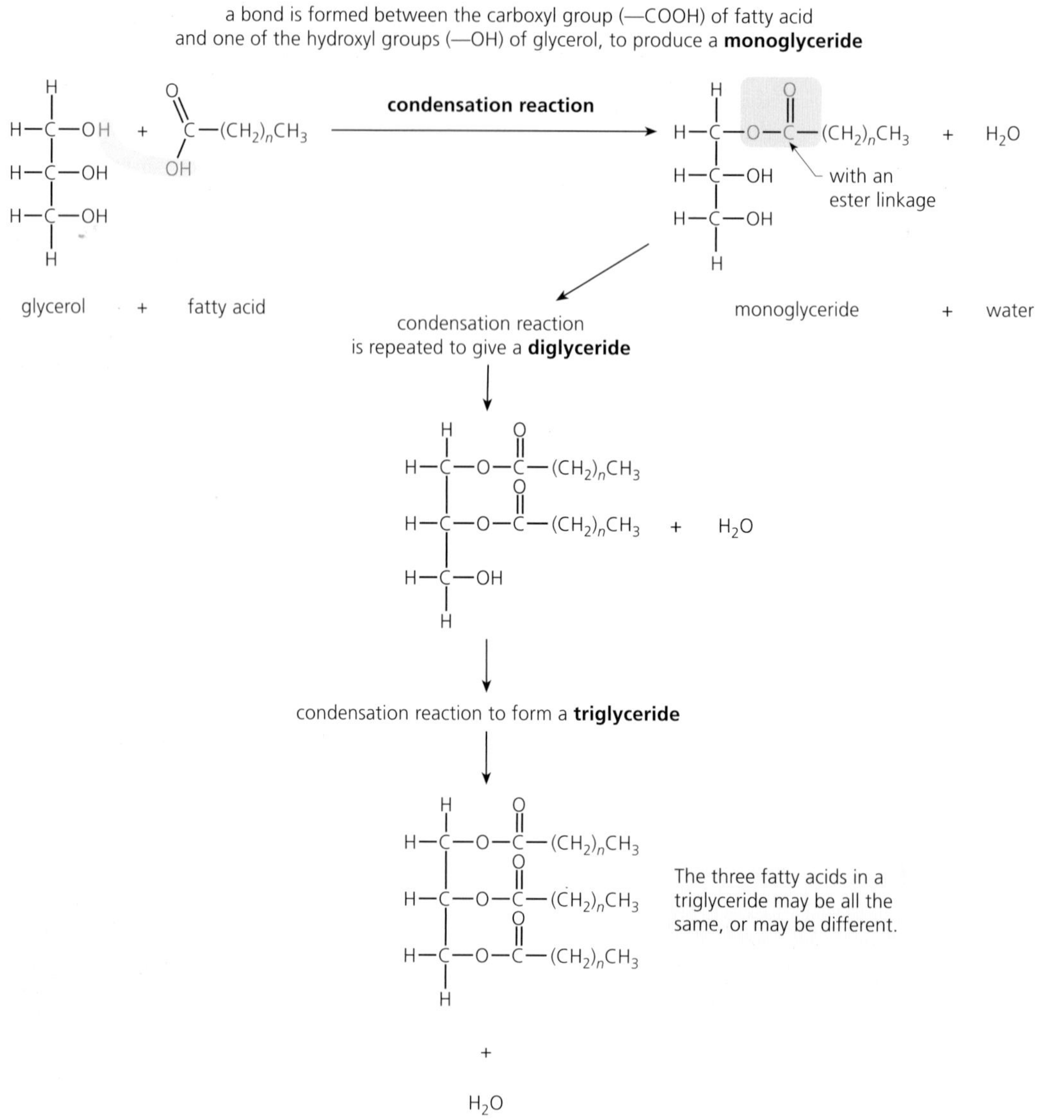

8 Distinguish between *condensation* and *hydrolysis reactions*. Give an example of each.

a bond is formed between the carboxyl group (—COOH) of fatty acid and one of the hydroxyl groups (—OH) of glycerol, to produce a **monoglyceride**

glycerol + fatty acid

condensation reaction

monoglyceride + water

with an ester linkage

condensation reaction is repeated to give a **diglyceride**

condensation reaction to form a **triglyceride**

The three fatty acids in a triglyceride may be all the same, or may be different.

Saturated and unsaturated lipids

We have seen that the different fatty acids combined in a triglyceride may vary in their length. In fact, the fatty acids present in dietary lipids (the lipids we commonly eat) vary in another, more important, way too. To understand this latter difference we need to note another property of carbon atoms and the ways they may combine together in chains. This difference concerns the existence of **double bonds** (Figure 1.10).

A double bond is formed when adjacent carbon atoms share *two pairs* of electrons, rather than the single electron pair shared in a single bond. Carbon compounds that contain double carbon–carbon bonds are known to chemists as **unsaturated** compounds. On the other hand, when all the carbon atoms in the hydrocarbon tail of an organic molecule are combined together by single bonds (the hydrocarbon chain consists of $-CH_2-CH_2-$ repeated again and again), then the compound is described as **saturated**.

This difference, illustrated in Figure 1.11, is of special importance in the fatty acids that are built into our dietary lipids.

■ Lipids built exclusively from saturated fatty acids are known as **saturated fats**. Examples of saturated fatty acids include palmitic acid (Figure 1.11) and stearic acid, both of which are major constituents of butter, lard, suet and cocoa butter. Butyric acid is present in only small amounts in milk fat and butter, but plays a major part in creating their characteristic flavours.

■ Lipids built from one or more unsaturated fatty acid are referred to as **unsaturated fats** by dieticians. Oleic acid (Figure 1.11) is an example, occurring in significant quantities in many common fats and oils, but in olive oil and rapeseed oil it makes up about 70% of the fatty acids present.

Figure 1.10
A carbon–carbon double bond.

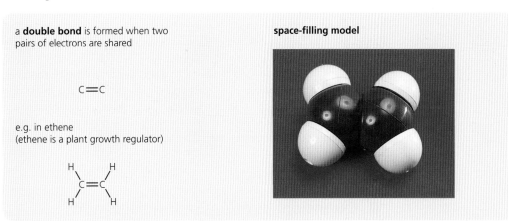

a **double bond** is formed when two pairs of electrons are shared

$$C = C$$

e.g. in ethene
(ethene is a plant growth regulator)

space-filling model

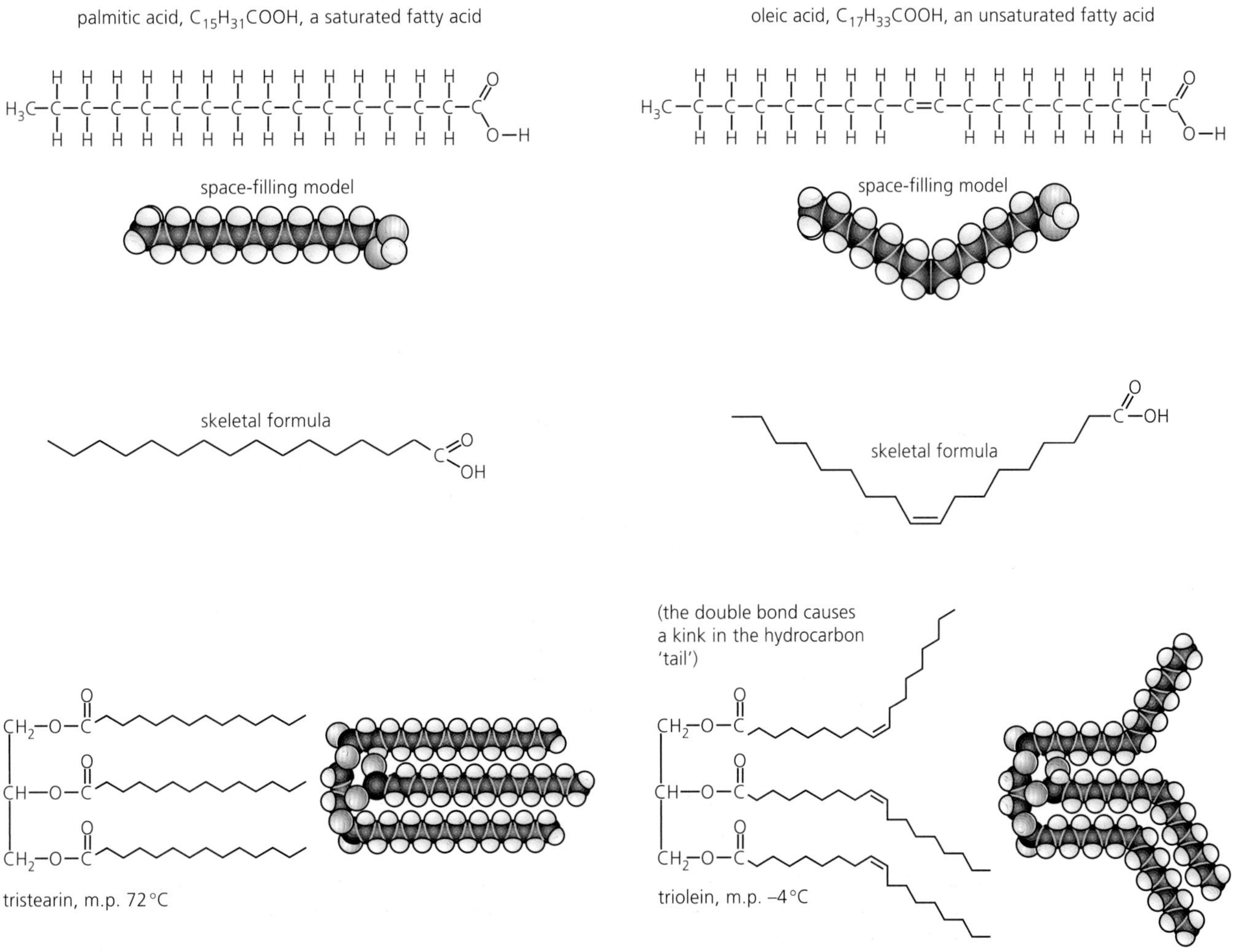

palmitic acid, $C_{15}H_{31}COOH$, a saturated fatty acid

oleic acid, $C_{17}H_{33}COOH$, an unsaturated fatty acid

space-filling model

space-filling model

skeletal formula

skeletal formula

(the double bond causes a kink in the hydrocarbon 'tail')

tristearin, m.p. 72 °C

triolein, m.p. –4 °C

Figure 1.11 Saturated and unsaturated fatty acids and triglycerides formed from them.

Finally, the term 'polyunsaturated fat' needs a note of explanation, although its relevance to the health and well-being of our arteries will be discussed later (page 28).

Where there is just one double bond in the carbon chain of a fatty acid, the compound is referred to as a **monounsaturated** fatty acid. However, it is possible and common for there to be two or more double bonds in the carbon chain. Linoleic acid, which has two double bonds, occurs in large amounts in vegetable seed oils, such as maize, soya and sunflower seed oils. Linolenic acid, which has three double bonds, occurs in small amounts in vegetable oils. These are examples of **polyunsaturated** fatty acids.

Fats with unsaturated fatty acids melt at a lower temperature than those with saturated fatty acids, because their kinked hydrocarbon tails do not allow the molecules to pack closely together (Figure 1.11), as they do in saturated fats. This difference between saturated and polyunsaturated fats is important in the manufacture of margarine and butter spreads – because they contain a proportion of polyunsaturated fats, they are soft enough to spread easily, even when used straight from the fridge.

■ Extension: Proteins and nucleic acids

The other major biochemical components of cells and organisms are the proteins and nucleic acids. These groups of organic compounds are discussed in Chapter 2.

■ 1.2 Internal transport in the body

Living cells require a supply of water and nutrients such as glucose and amino acids, and most need oxygen. The waste products of cellular metabolism have to be removed, too. In single-celled organisms and very small organisms, internal distances are small, so here movements of nutrients can occur efficiently by **diffusion** (Chapter 2), although some substances require to be transported across membranes by **active transport**. In larger organisms, these mechanisms alone are insufficient – an internal transport system is also required to service the needs of the cells.

Internal transport systems at work are examples of **mass flow**. In mass flow, fluid moves in response to a pressure gradient, flowing from a region of high pressure to regions of lower pressure. Carried in the fluid are any suspended and dissolved substances present.

The more active an organism, the higher the rate at which nutrients are required by its cells, and so the greater the need for an efficient internal transport system. Larger animals have a **blood circulatory system** that links the parts of the body and makes resources available where they are required.

Transport in mammals

Mammals have a **closed circulation** in which blood is pumped by a powerful, muscular heart and circulated in a continuous system of tubes – the **arteries**, **veins** and **capillaries** – under pressure. The heart has four chambers, and is divided into right and left sides. Blood flows from the right side of the heart to the lungs, where it is oxygenated, and then back to the left side of the heart. From here it is pumped around the rest of the body and back to the right side of the heart. As the blood passes twice through the heart in every single circulation of the body, this is called a **double circulation**.

The circulatory system of mammals is shown in Figure 1.12, alongside alternative systems. Looking at these other systems helps us to understand the features of the mammalian circulation.

Fish also have a closed circulation, but here blood flows only once through the heart in every circulation of the body, a condition known as a **single circulation**. This means that the blood pressure lost in passing through the gills is not re-established before the oxygenated blood is distributed to the respiring tissues.

Insects have an **open circulation** in which blood is pumped out by a long, tubular heart into spaces in the body cavity called sinuses, where it bathes the body organs directly, but under very low pressure. From the body organs, the blood re-enters the heart through openings controlled by valves, and it is re-circulated. At the same time, in many insects, air is delivered directly to the respiring tissues using a system of tiny, branching tubes called tracheae.

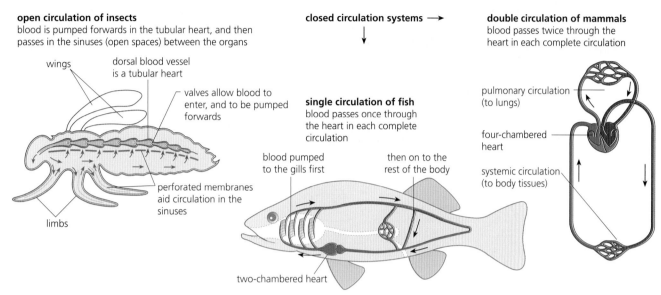

Figure 1.12 Open and closed circulations.

9 In an open circulatory system, there is 'little control over circulation'. What does this mean?

It becomes clear that the major advantages of the mammalian circulation are that:

- oxygenated blood is delivered at high pressure to all regions of the body simultaneously
- oxygenated blood reaches the respiring tissues undiluted by deoxygenated blood.

In discussing the mammalian blood circulation, we will take the human circulation as the example.

The transport medium – the blood

Blood is a special tissue consisting of a liquid medium called **plasma** in which are suspended red cells (**erythrocytes**), white cells (**leucocytes**), and the **platelets** (Figure 1.13). The plasma is the medium for exchange of substances between cells and tissues; the red cells are involved in transport of respiratory gases; and the white cells are adapted to combat infection. The roles of the components of blood are summarised in Table 1.3.

Table 1.3 The components of the blood and their roles.

Component	Role
plasma (note: 'serum' is plasma from which all cells and the soluble protein fibrinogen have been removed)	transport of: ■ nutrients from gut or liver to all cells ■ excretory products e.g. urea from the liver to the kidneys ■ hormones from the endocrine glands to all tissues and organs ■ dissolved proteins that have roles including regulating the osmotic concentration (water potential) of the blood ■ dissolved proteins that are antibodies ■ heat to all tissues
red cells	transport of: ■ oxygen from the lungs to respiring cells ■ carbon dioxide from respiring cells to the lungs (also carried in the plasma)
white cells	lymphocytes have major roles in the immune system, including forming antibodies phagocytes ingest bacteria or cell fragments
platelets	involved in the blood clotting mechanism (see below)

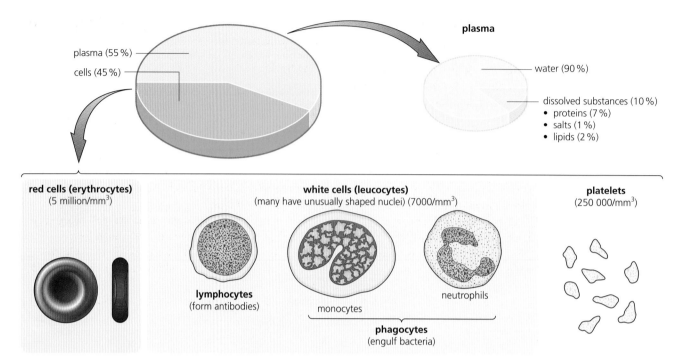

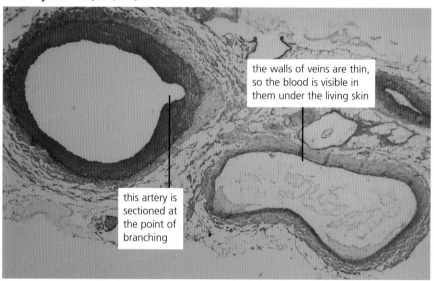

Figure 1.13 The composition of the blood.

TS artery and vein, LP (×20)

In sectioned material (as here), veins are more likely to appear squashed, whereas arteries are circular in section.

the walls of veins are thin, so the blood is visible in them under the living skin

this artery is sectioned at the point of branching

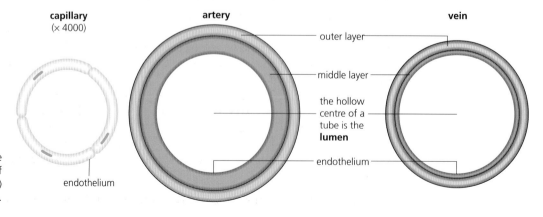

Figure 1.14 The structure of the walls of arteries (x20), veins (x20) and capillaries (x4000).

The plumbing of the circulatory system – arteries, veins and capillaries

There are three types of vessel in the circulatory system:

- **arteries**, which carry blood away from the heart
- **veins**, which carry blood back to the heart
- **capillaries**, which are fine networks of tiny tubes linking arteries and veins.

Both arteries and veins have strong, elastic walls, but the walls of the arteries are very much thicker and stronger than those of the veins. The strength of the walls comes from the collagen fibres present, and the elasticity is due to the elastic fibres and involuntary (smooth) muscle fibres. The walls of the capillaries, on the other hand, consist of endothelium only (endothelium is the innermost lining layer of arteries and veins). Capillaries branch profusely and bring the blood circulation close to cells – no cell is far from a capillary.

Blood leaving the heart is under high pressure, and travels in waves or **pulses**, following each heart beat. By the time the blood has reached the capillaries, it is under very much lower pressure, without a pulse. This difference in blood pressure accounts for the differences in the walls of arteries and veins. Figure 1.14 shows an artery, vein and capillary vessel in section, and details of the wall structure of these three vessels.

Because of the low pressure in veins, there is a possibility of backflow here. Veins have **valves** at intervals, which prevent this (Figure 1.15).

Figure 1.15 The valves in veins.

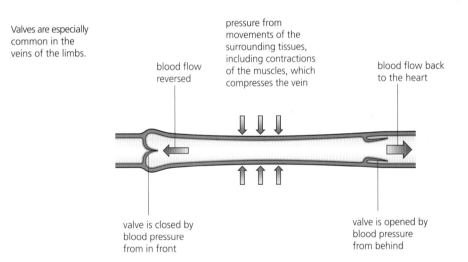

Valves are especially common in the veins of the limbs.

pressure from movements of the surrounding tissues, including contractions of the muscles, which compresses the vein

blood flow reversed

blood flow back to the heart

valve is closed by blood pressure from in front

valve is opened by blood pressure from behind

The arrangement of arteries and veins

We have already noted that mammals have a double circulation. It is the role of the right side of the heart to pump deoxygenated blood to the lungs. The arteries, veins and capillaries serving the lungs are known as the **pulmonary circulation**. The left side of the heart pumps oxygenated blood to the rest of the body. The arteries, veins and capillaries serving the body are known as the **systemic circulation**.

In the systemic circulation, organs are supplied with blood by an artery branching from the main artery known as the **aorta**. Within individual organs, the arteries branch into numerous arterioles (smaller arteries), and the smallest arterioles supply the capillary networks. Capillaries drain into venules (smaller veins), and venules join to form veins. The veins join the main vein (**vena cava**) carrying blood back to the heart. The branching sequence in the circulation is, therefore:

aorta → artery → arteriole → capillary → venule → vein → vena cava

Activity 1.4: Designing a summary table – arteries, capillaries and veins

Arteries and veins are often named after the organs they serve (Figure 1.16). The blood supply to the liver is via the hepatic artery, but the liver also receives blood directly from the small intestine, via a vein called the **hepatic portal vein**. This brings much of the products of digestion, after they have been absorbed into the blood circulation in the gut.

Figure 1.16 The layout of the human circulation.

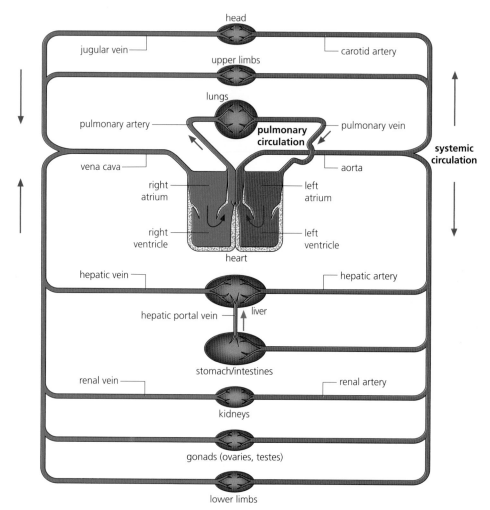

The circulatory system as shown here is simplified, e.g. limbs, lungs, kidneys and gonads are paired structures in the body.

The blood clotting mechanism

If the event of a break in our closed blood circulation, the danger of a loss of blood and the possibility of a fall in blood pressure arise. It is by the clotting of blood that escapes are prevented, either at small internal haemorrhages, or at cuts and other wounds. In these circumstances, a clot both stops the outflow of blood and reduces invasion opportunities for disease-causing organisms (pathogens). Subsequently, repair of the damaged tissues can get underway. Initial conditions at the wound trigger a **cascade of events** by which a blood clot is formed.

What is meant by a 'cascade' of events?

Firstly, **platelets** collect at the site. These are components of the blood that are formed in the bone marrow along with the red and white cells, and are circulated throughout the body, suspended in the plasma, with the blood cells. Platelets are actually cell fragments, disc-shaped and very small (only 2 μm in diameter) – too small to contain a nucleus. Each platelet consists of a sac of cytoplasm rich in vesicles containing enzymes, and is surrounded by a plasma membrane. Platelets stick to the damaged tissues and clump together there. (At this point they change shape from sacs to flattened discs with tiny projections that interlock.) This action alone seals off the smallest breaks.

The collecting platelets release a **clotting factor** (a protein called thromboplastin), which is also released by damaged tissues at the site. This clotting factor, along with vitamin K and calcium ions, always present in the plasma, causes a soluble plasma protein called **prothrombin** to be converted to the active, proteolytic enzyme **thrombin**. The action of this enzyme is to convert another soluble blood protein, **fibrinogen**, to insoluble **fibrin** fibres at the site of the cut. Within this mass of fibres, red cells are trapped, and the blood clot has formed.

Figure 1.17 The blood clotting mechanism.

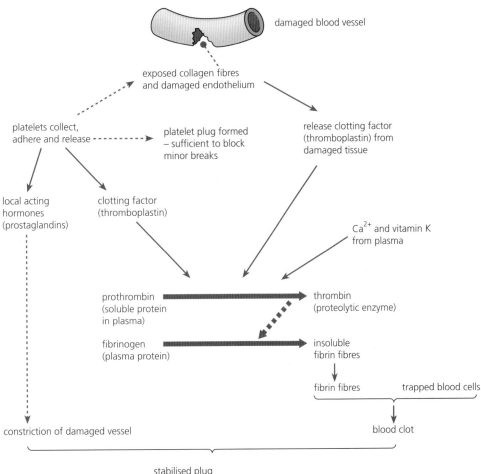

SEM of blood clot
showing meshwork of fibrin fibres and trapped blood cells

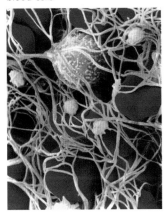

10 Identify the correct sequence of the following events during blood clotting: fibrin formation; clotting factor release; thrombin formation.

It is most fortunate that clot formation is not normally activated in the intact circulation; clotting is triggered by the abnormal conditions at the break. The complex sequence of steps involved in clotting may be seen as an essential **fail-safe mechanism**. This is necessary because a casual formation of a blood clot within the intact circulatory system immediately generates the risk of a dangerous and possibly fatal blockage in the lungs, heart muscle or brain (page 25).

The heart as a pump

The human heart is the size of a clenched fist. It is found in the thorax between the lungs and beneath the breast bone (sternum). The heart is a hollow organ with a muscular wall, and is contained in a tightly fitting membrane, the pericardium – a strong, non-elastic sac that anchors the heart within the thorax.

The cavity of the heart is divided into four chambers, with those on the right side of the heart completely separate from the left side. The two upper chambers are thin-walled **atria** (singular: atrium). These receive blood into the heart. The two lower chambers are thick-walled

Figure 1.18 The structure of the heart.

heart viewed from the front of the body with pericardium removed

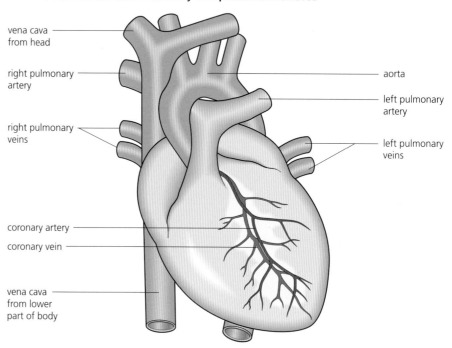

heart in LS

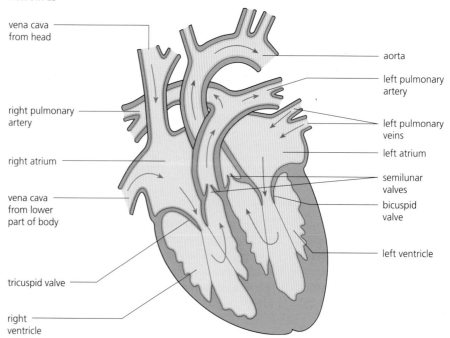

ventricles. The ventricles pump blood out of the heart, with the muscular wall of the left ventricle much thicker than that of the right ventricle. However, the volumes of the right and left sides (the quantities of blood they contain) are identical.

Note that the walls of the heart (the heart muscle) are supplied with oxygenated blood via **coronary arteries**. These arteries, and the capillaries they serve, deliver to the heart muscle fibres the **oxygen** and **nutrients** essential for the maintenance of the pumping action.

The **valves** of the heart prevent backflow of the blood, thereby maintaining the direction of flow through the heart. The **atrio-ventricular valves** are large valves, positioned to prevent backflow from ventricles to atria. The edges of these valves are supported by tendons anchored to the muscle walls of the ventricles below. The valves on the right and left sides of the heart are individually named: on the right side, the **tricuspid valve**; on the left, the **bicuspid** or mitral valve.

A different type of valve separates the ventricles from pulmonary artery (right side) and aorta (left side). These are pocket-like structures called **semilunar valves**, rather similar to the valves seen in veins. These cut out backflow from aorta and pulmonary artery into the ventricles, as the ventricles relax between heart beats.

11 The edges of the atrio-ventricular valves have non-elastic strands attached, which are anchored to the ventricle walls (Figure 1.18). Suggest the role of these strands.

Figure 1.19 The cardiac cycle.

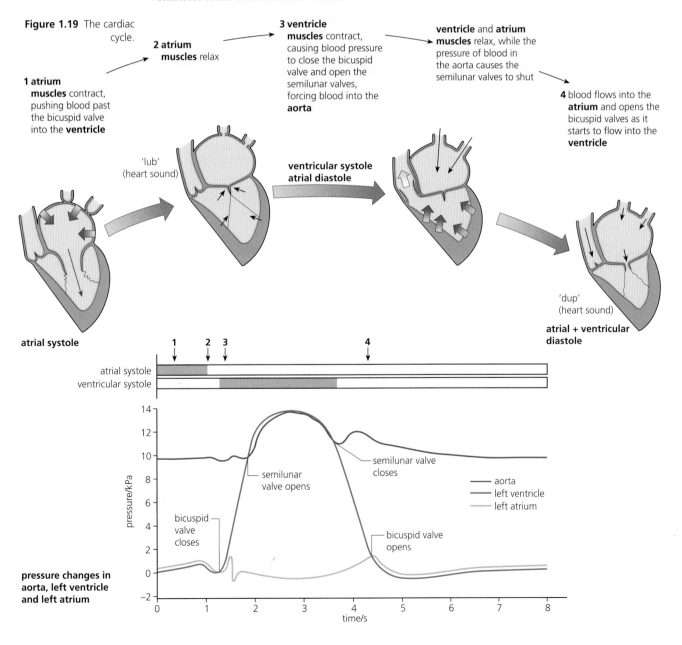

1 atrium muscles contract, pushing blood past the bicuspid valve into the **ventricle**

2 atrium muscles relax

3 ventricle muscles contract, causing blood pressure to close the bicuspid valve and open the semilunar valves, forcing blood into the **aorta**

ventricle and **atrium muscles** relax, while the pressure of blood in the aorta causes the semilunar valves to shut

4 blood flows into the **atrium** and opens the bicuspid valves as it starts to flow into the **ventricle**

'lub' (heart sound)

ventricular systole atrial diastole

'dup' (heart sound)

atrial + ventricular diastole

atrial systole

pressure changes in aorta, left ventricle and left atrium

Activity 1.5: Simulation of the cardiac cycle

Activity 1.6: Wider reading – 'The heart'

A* Extension 1.2: Origin of the heart beat

The action of the heart – the cardiac cycle

The heart normally beats about 75 times per minute – approximately 0.8 seconds per beat. In each beat, the heart muscle contracts strongly, and this is followed by a period of relaxation. As the muscular walls of a chamber of the heart contract, the volume of that chamber decreases. This increases the pressure on the blood contained there, forcing the blood to a region where pressure is lower. Since the valves prevent blood flowing backwards, blood consistently flows on through the heart.

Look at the steps involved in contraction and relaxation, illustrated in the left side of the heart, in Figure 1.19. (Both sides function together, in exactly the same way, of course.)

We start at the point where the atrium contracts. Blood is pushed into the ventricles (where the contents are under low pressure) by contraction of the walls of the atrium. This contraction also prevents backflow by blocking off the veins which brought the blood to the heart. This contraction step is known as **atrial systole**.

The atrium now relaxes. The relaxation step is called **atrial diastole**.

Next the ventricle contracts, and contraction of the ventricle is very forceful indeed. This step is known as **ventricular systole**. The high pressure this generates slams shut the atrio-ventricular valve and opens the semilunar valves, forcing blood into the aorta. A **pulse**, detectable in arteries all over the body, is generated.

This is followed by relaxation of the ventricles. Each contraction of cardiac muscle is followed by relaxation and elastic recoil. This stage is referred to as **ventricular diastole**.

12 Examine the data on pressure change during the cardiac cycle in Figure 1.19. Determine or suggest why:
   a pressure in the aorta is always significantly higher than that in the atria
   b pressure falls most abruptly in the atrium once ventricular systole is underway
   c the semilunar valve in the aorta does not open immediately ventricular systole commences
   d when ventricular diastole commences, there is a significant delay before the bicuspid valve opens, despite rising pressure in the atrium
   e it is significant that about 50% of the cardiac cycle is given over to diastole.

Blood pressure and its measurement

By blood pressure we mean the pressure of the blood flowing through the arteries.

We have noted that, initially, flow is a surge or **pulse**, but pressure falls as blood flows on through the smaller arteries, arterioles and capillaries to the veins. Pulsation has entirely disappeared by the time the capillaries have been reached (Figure 1.20). It is the resistance the

Figure 1.20 Changing blood pressure in the circulatory system.

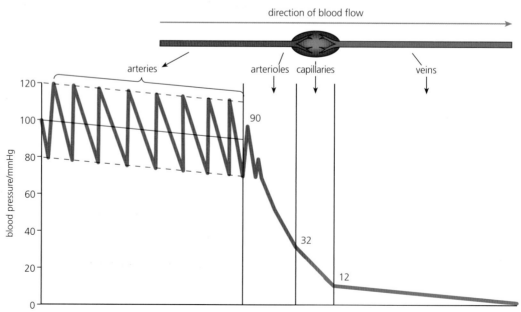

blood experiences that slows down the flow and causes the blood pressure to fall. This is described as peripheral resistance.

So it is arterial blood pressure that is measured, in a part of our body relatively close to the heart. Blood pressure is quoted as two values (typically, one over the other). The high pressure is produced by ventricular systole (systolic pressure), and is followed by low pressure at the end of ventricular diastole (diastolic pressure) (Table 1.4). Normally, systolic and diastolic pressures are about 15.8 and 10.5 kPa respectively.

Note, though, that the medical profession gives these values as 120 and 70–80 mmHg (see Figure 1.21 – the term 'mmHg' was recognised as a unit of pressure before SI units were introduced).

To measure these two values, an instrument called a sphygmomanometer is used with a stethoscope, as shown in Figure 1.21.

1   The cuff is inflated and blood flow is monitored in the arm (brachial) artery at the elbow, using the stethoscope. Inflation is continued until there is no sound (that is, no flow of blood, because the cuff has completely occluded the artery).
2   Air is now allowed to escape from the cuff, slowly, until blood can just be heard spurting through the constriction point in the artery. This pressure at the cuff is recorded, for it is equal to the maximum pressure created by the heart (**systolic pressure**).
3   Pressure in the cuff is allowed to drop until the blood can be heard flowing constantly. This is the lowest pressure the blood falls to between beats (**diastolic pressure**).

The pascal (Pa) and its multiple the kilopascal (kPa) are generally used by scientists to measure pressure, but in medicine the older unit of pressure, 'millimetre of mercury' (mmHg) is still used (1 mmHg = 0.13 kPa).

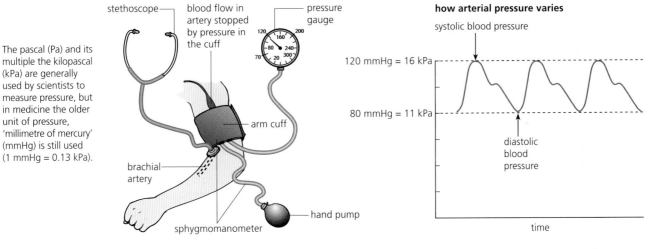

Figure 1.21 Measuring blood pressure.

Table 1.4 Screening of blood pressure in adults – a range of conditions.

Systolic	Diastolic	Condition	Response
120	80	optimum	
<130	<85	normal	biennial checks
130–139	85–89	high–normal	annual checks
140–159	90–99	stage 1 hypertension	check in 2 months
160–179	100–109	moderate (stage 2) hypertension	treatment essential if these conditions persist
180–209	110–119	severe (stage 3) hypertension	
> 210	120	very severe (stage 4) hypertension	

Activity 1.7: Measuring blood pressure – alternatives

■ Investigation: heart rate – a physiological experiment

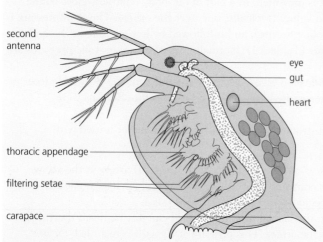

Figure 1.22 *Daphnia*, seen from the side.

Animal physiology questions how and why the body parts of organisms function as they do. It is an experimental science. The outcomes often have applications for exercise physiology and sports science, and all activities where bodies may have to function under extreme conditions and up to the limits of their potential. Equally, the physiology of normal, healthy bodies is the standard by which the impact of disease and conditions of malfunction are assessed by the medical and veterinary professions and the pharmaceutical industry.

Sometimes experimental data are obtained from experiments with human subjects, sometimes from studies with other animals, and sometimes from isolated tissue studies. **The nature of ethics is discussed in Chapter 2 (pages 95–99), and the ethics of animal experimentation are considered in *How science works* activity 1.2, below.**

Here we introduce experimental physiology by considering a simple, practical study of the effect of temperature on the heart rate of *Daphnia* (a water flea), as it might be conducted in a school or college laboratory. *Daphnia* is a small aquatic animal of freshwater habitats (Figure 1.22). It does not regulate its internal body temperature, as mammals do.

Aim: To measure the heart rate of *Daphnia* at different temperatures.

Method: Transfer a large *Daphnia* from a culture at known temperature to a cavity microscope slide, using a paint brush. Remove the excess liquid with a filter paper, so as to cause the animal to lie on its side.

Place the slide on the stage of the microscope. Locate and examine the *Daphnia* using the low-power objective. The heart will be observed, beating extremely fast.

Working in pairs, practise counting heart beats by setting up a basic calculator to add one digit repetitively (key in 1++, then press the = key in time with the heart beat) for 20-second periods, measured using a stopwatch. (See also video capture of data, in HSW criterion 3, below.) Record the results of three consecutive counting periods and calculate the mean value, expressed as counts per minute.

Using identical culture solutions at temperatures of 5, 10, 15, 20, 25 and 30 °C, place the *Daphnia* in these cultures in turn, allowing 5 minutes to elapse before taking a new heart rate determination.

Data presentation: Present the results in a table and graph to show the relationship between the rate of the heart beat and the temperature of the water.

Discussion and conclusion: What relationship between the heart rate and the temperature of water is shown?

Evaluation: The heart beat of this animal might be expected to increase with external temperature. Whatever these results indicate, what are the sources of error and inaccuracy in this experiment?

Alternative studies: This technique could be adapted to investigate the effects of drugs like caffeine or ethanol, when applied at appropriately dilute concentrations.

HSW 1.2: Criteria 2, 3, 4, 5, 8 and 10 – The effects of temperature on heart rate in *Daphnia*

Activity 1.8: Listening to heart sounds and taking the pulse

Activity 1.9: Studies of changing pulse rates

A* Extension 1.3: The electrocardiogram and its diagnostic importance

Exchange in the tissues – formation of tissue fluid

The formation of tissue fluid assists in the delivery of nutrients to cells, and the removal of waste products. Tissue fluid is formed from the plasma, components of which escape from the blood and pass between the cells in most of the tissues of the body. Red cells and most of the blood proteins are retained in the capillaries.

The walls of the capillaries are selectively permeable to many components of the blood plasma, including glucose and mineral ions. Nutrients like these, in low concentration in the tissues, diffuse from the plasma (Figure 1.23). There are also tiny gaps in the capillary walls, found to vary in size in different parts of the body, which facilitate formation of tissue fluid. It is the pressure of the blood that drives fluid out (that is, the hydrostatic pressure generated as the heart beats). Meanwhile, the proteins and some other components are retained in the blood. These soluble substances maintain an osmotic gradient, so some of the water forced out by hydrostatic pressure returns to the blood by osmosis all along the capillary. However, initially there is a net outflow because the hydrostatic pressure is greater than fluid movement due to the osmotic gradient.

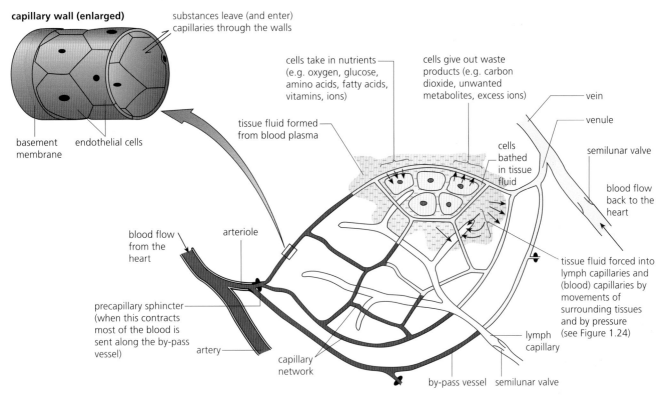

Figure 1.23 Exchange between blood and cells via tissue fluid.

Return of tissue fluid to the circulation
Further along the capillary, there is a net inflow of tissue fluid to the capillary (Figure 1.24). Hydrostatic pressure has now fallen as fluid is lost from the capillaries. Water returns by osmosis, and a diffusion gradient carries unused metabolites and excretory material back into the blood.

Figure 1.24 Forces for exchange in capillaries.

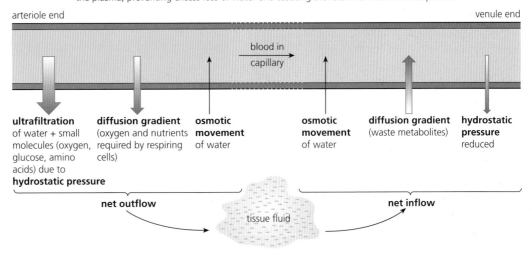

13 What components of the blood are not found in tissue fluid?

Not all tissue fluid returns to the blood capillaries – some enters the lymph capillaries. Molecules too large to enter blood capillaries can pass into the lymph system at tiny valves in the vessel walls. Liquid is moved along these capillaries by compression due to body movements, and backflow is prevented by valves. Lymph finally drains back into the blood circulation in veins close to the heart.

> ### ■ Extension: Oedema
>
> If formation of tissue fluid greatly exceeds reabsorption, the result is a much increased volume of tissue fluid, which results in swelling (that is, oedema). This condition arises due to either increased blood pressure (page 20), or increased permeability of capillary walls.

■ 1.3 Disease of the human circulatory system

Diseases of the heart and blood vessels are known as **cardiovascular diseases** (**CVD**). These are responsible for more premature deaths in the developed world than any other single cause. By premature death, we mean a death that occurs before the age of 75 years. Most of these are due to **atherosclerosis** – the progressive degeneration of the artery walls. The structure of a healthy artery wall is shown in Figure 1.14 (page 15).

Remind yourself of this structure now.
The steps in the development of an atherosclerotic condition are as follows.

1 **Endothelial damage**. Healthy arteries have pale, smooth linings, and the walls are elastic and flexible. However, in arteries that have become unhealthy, the walls have strands of yellow fat deposited under the endothelium. This fat builds up from certain lipoproteins and from cholesterol that may be circulating in the blood. With the fatty streaks, fibrous tissue is laid down as well.

2 **Raised blood pressure**. These deposits start to impede blood flow, and contribute to raised blood pressure. Thickening of the artery wall leads to loss of elasticity, and this, too, contributes to raised blood pressure. In the special case of the arteries serving the heart muscle, the coronary arteries, progressive reduction of the blood flow impairs oxygenation of the cardiac muscle fibres, leading to chest pains, known as **angina**, usually brought on by physical exertions.

healthy

endothelium ——

flow of blood

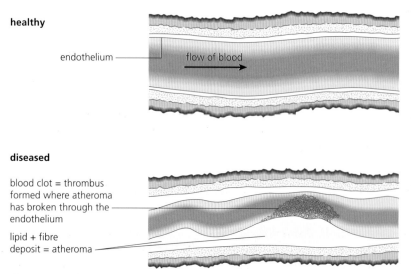

diseased

blood clot = thrombus formed where atheroma has broken through the endothelium ——

lipid + fibre deposit = atheroma ——

photomicrograph of diseased human artery in **TS** (×20)

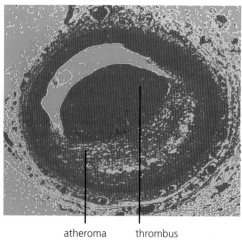

atheroma   thrombus

Figure 1.25 Atherosclerosis, leading to a thrombus.

14 Create a flow diagram of the sequence of changes occurring at a site in an ageing artery wall where, starting from a healthy condition, a thrombus forms and then departs as an embolus.

3 **Lesion formation and an inflammatory response**. Where the smooth lining actually breaks down, the circulating blood is exposed to the fatty, fibrous deposits. These lesions are known as atherosclerotic plaques. Further deposition occurs as cholesterol and triglycerides accumulate, and smooth muscle fibres and collagen fibres proliferate in the plaque. Blood platelets tend to collect at the exposed roughened surface, and these platelets release factors that trigger a defensive response called inflammation that includes blood clotting (see Figure 1.17, page 17). A blood clot may form within the vessel. This clot is known as a **thrombus**, at least until it breaks free and is circulated in the blood stream, whereupon it is called an **embolus**.

Myocardial infarctions, strokes and aneurysms

An embolus may be swept into a small artery or arteriole which is narrower than the diameter of the clot, causing a blockage. Immediately, the blood supply to the tissue downstream of the block is deprived of oxygen. Without oxygen, the tissue dies.

The arteries supplying the heart are the coronary arteries. These arteries are especially vulnerable, particularly those to the left ventricle. When heart muscle dies in this way, the heart may cease to be an effective pump. We say a heart attack has occurred (known as a **myocardial infarction**). Coronary arteries that have been damaged can be surgically by-passed (Figure 1.26).

Figure 1.26 A heart by-pass.

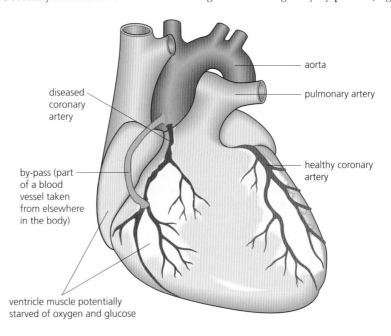

diseased coronary artery

aorta

pulmonary artery

healthy coronary artery

by-pass (part of a blood vessel taken from elsewhere in the body)

ventricle muscle potentially starved of oxygen and glucose

15 Identify the advantage and the disadvantage to the body of the tendency of platelets to release clotting factors at the site of damaged artery walls.

When an embolus blocks an artery in the brain, a **stroke** occurs. Neurones of the brain depend on a continuous supply of blood for oxygen and glucose. Within a few minutes of the blood supply being lost, the neurones affected will die. Neurones cannot be replaced, so the result of the blockage is a loss of some body functions controlled by that region of the brain.

In arteries where the wall has been weakened by atherosclerosis, the remaining layers may be stretched and bulge under the pressure of the blood pulses. Ballooning of the wall like this is called an **aneurysm**. An aneurysm may burst at any time.

A* Extension 1.4: Aortic aneurysm

Activity 1.10: Wider reading – 'Coronary heart disease: The number one killer'

Factors affecting the incidence of coronary heart disease (CHD)

In the developed world, cardiovascular diseases remain high on the list of most serious health problems, despite recent improvements. For example, in the UK, CVD is responsible for the second highest percentage of deaths before the age of 75 years in both males and females (Figure 1.27). Of these deaths, the majority are due to CHD.

Males under 75 years   **Females under 75 years**

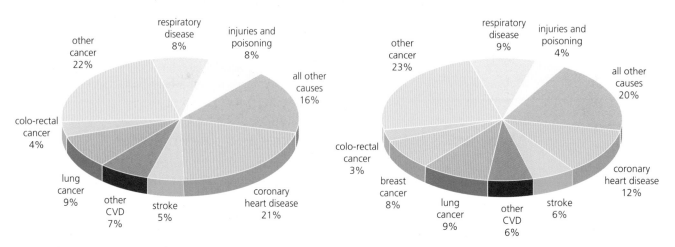

Figure 1.27 Premature death by causes in the UK, 2004.

Another feature of CHD is that the condition may go largely unnoticed for many years – often until it is too late.

What factors contribute to the high incidence of coronary heart disease?

Firstly, there are certain major risk factors that cannot be changed, as described below.

■ **Increasing age**. The risk of coronary atherosclerosis increases with age. Over 80% of people who die from CHD are aged 65 years or more. However, a word of caution is required here, for the evidence is that the genesis of this condition may lay much earlier in life. Post mortem studies of soldiers killed in action have disclosed early but well-developed 'fatty plaques' in the arteries of young males of average age 22 years, suggesting that these commenced development in their bodies during their adolescence. This suggests why, for individuals brought up in the developed world, after the ages of 35 in men and 45 in women, the chance of dying from CHD increases dramatically.

■ **Genetic factors** (including ethnicity). The occurrence of heart attacks at a relatively early age frequently runs in families, suggesting that there are genes that may confer vulnerability to CHD. Children of parents with heart diseases are distinctly more likely to develop the condition, too. This aspect will become clearer as knowledge of the human genome is developed to identify the roles of individual genes and their effects on metabolism. Additionally, some races are more prone to CHD and strokes. This is the case with Afro-Caribbean people, for example. Another genetic factor beyond dispute is gender, for the possession of a Y chromosome (being a male) predisposes to greater risk of CHD than that carried by females.

Secondly, there are major risk factors that we *can* control or reduce, either by change in lifestyle, or by medical treatments, or both. These are described here.

■ **Hypertension**. Persistently high blood pressure is defined as systolic pressure greater than 140 mmHg and diastolic pressure greater than 90 mmHg. Remember, a blood pressure of 120/80 is normal and preferable in an adult (Table 1.4, page 21).

Hypertension is known as a 'silent killer' because of the damage it does to the heart by increasing its workload, causing the heart to enlarge and weaken with time. It also causes damage to blood vessels, accelerating onset of atherosclerosis. The brain and kidneys are also damaged, although without causing noticeable discomfort. Hypertension increases the risk of strokes, too, and it makes a brain haemorrhage more likely.

The following factors in this list all make the condition of hypertension more likely. However, hypertension is a condition that, once detected and regularly monitored, may be successfully treated with drugs (see below).

- **Smoking.** The habit of cigarette smoking generates the greatest risk of fatal ill-health, especially from cardiovascular diseases. It is principally the nicotine and carbon monoxide in tobacco smoke that damage the cardiovascular system. Carbon monoxide combines irreversibly with the pigment haemoglobin in red cells, reducing the ability of the blood to transport oxygen to all respiring cells, including to cardiac muscle fibres. The effects of nicotine are via its stimulation of adrenaline production. This hormone triggers an increase in heart rate and causes arteries to contract (**vasoconstriction**). The result is raised blood pressure.

There have been huge investments in the tobacco industry via the growth of the crop and the manufacturing of tobacco products, for a long time. The enthusiastic endorsement of cigarette smoking in the developed world (manifested by the free availability of cigarettes to the troops in two World Wars, for example) has now spread to developing countries as fresh markets for tobacco products have opened up. Consequently, the dangers of smoking have not had the attention they deserve, and some people have even suggested the evidence against tobacco is equivocal.

Yet from the earliest statistical studies there has been no room for doubts. The late Dr Richard Doll and colleagues, working at St Thomas' Hospital, London from 1947, investigated the cause of death of a sample of 3500 people admitted to hospital for treatment. Whatever the symptoms these people had, the vast majority of those who were smokers later died from a cardiovascular disease.

This study was followed up by one of a group of 40 000 healthy, working doctors (among whom the habit of smoking was widespread at that time) which was more conclusive still. Some of Richard Doll's data from his studies with doctors are shown in Table 1.5. Today, few doctors smoke.

Table 1.5 Mortality from cardiovascular disease caused by cigarette smoking.

Cause of death	Non-smokers	Continuing cigarette smokers
CHD	606	2067
stroke	245	802
aneurysm	14	136
atherosclerosis	23	111
Total	**888**	**3116**

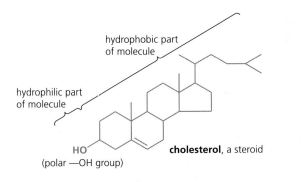
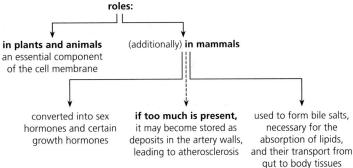

Figure 1.28 The steroid cholesterol.

■ **High levels of cholesterol in the blood.** Lipids are a more diverse group of biochemicals than just the triglycerides we met earlier in this chapter (page 9), for they include steroids. Steroids occur widely in nature, and one very important steroid is **cholesterol** (Figure 1.28). The 'skeleton' of a steroid is a set of complex rings of carbon atoms. The bulk of the molecule is hydrophobic but the polar –OH group is hydrophilic.

Lipids are absorbed into the body in the gut, and are stored as body fat. They have to be transported about the body, both in their role as respiratory substrates for the transfer of energy, and as cholesterol for use in the production of steroid hormones and the maintenance and repair of cell membranes. Since they are insoluble in water, they are carried in association with proteins, as either **low-** or **high-density lipoproteins** (LDLs or HDLs) according to the relative proportions of protein and lipid. Triglycerides combine with them. These components are introduced in Table 1.6.

Table 1.6 Low- and high-density lipoproteins.

	Protein (raises density)/%	Lipid (reduces density)/%	Particle size/nm*	Known as
low-density lipoprotein (LDL)	10–27	5–61	20–90	saturated fats combine, too, and they are collectively known as 'bad cholesterol'
high-density lipoprotein (HDL)	50	3	7–10	unsaturated fats combine, too, and they are collectively known as 'good cholesterol'

* A nanometre is an SI-derived unit of length, and is 10^{-9} metres.
So, a metre is divided into 1000 millimetres, a millimetre is divided into 1000 micrometres (μm or microns), and a micrometre is divided into 1000 nanometres (nm) – a really small unit of length!

Most cholesterol is transported as LDLs, but an excess of these in the blood stream has been shown to block up the many receptor points in the cell membranes of cells that metabolise or store lipid, leaving even higher quantities of LDLs circulating in the blood plasma. The excess is then deposited under the endothelium of artery walls, beginning or enhancing plaque formations. However, monosaturated fats help remove the circulating LDLs, and polyunsaturated fats are even more beneficial for they further increase the efficiency of the receptor sites at removing 'bad cholesterol' from the blood.

A note of caution is needed here. While we can (and should) avoid a diet that is excessively rich in saturated fats and cholesterol, this lipid is an essential body metabolite which is manufactured in the liver in the absence of absorbed dietary cholesterol. To some extent, our blood cholesterol levels are genetically controlled.

Figure 1.29 The relationship between deaths from CHD and blood serum cholesterol levels.

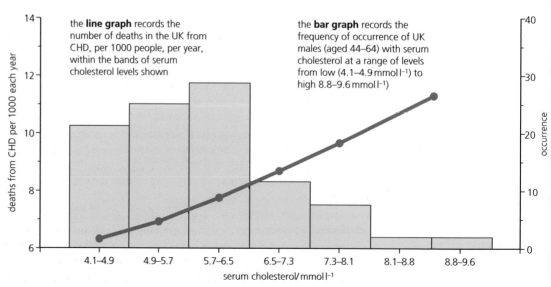

the **line graph** records the number of deaths in the UK from CHD, per 1000 people, per year, within the bands of serum cholesterol levels shown

the **bar graph** records the frequency of occurrence of UK males (aged 44–64) with serum cholesterol at a range of levels from low (4.1–4.9 mmol l⁻¹) to high 8.8–9.6 mmol l⁻¹)

A* Extension 1.5:
Cholesterol home test
kits – a critical review

What is the evidence for a causal relationship between blood LDLs and CHD?

A causal relationship is suggested by statistical studies of deaths from CHD per 1000 of the population each year, plotted against the levels of cholesterol and LDLs measured in blood (serum) (Figure 1.29). The establishment of the actual role of LDLs in triggering CHD is provided by experimental laboratory and clinical evidence that destructive plaques are created as a result of raised levels of blood serum LDLs, as described above.

■ **Alcohol.** An excessive intake of alcohol leads to raised blood pressure, to damaged heart muscle and to irregular heart beats. It also causes raised LDL levels in the blood, and is associated with certain cancers, too. This harm may arise when more than one unit of alcohol a day is imbibed by women and more than two units per day by men. In those who consistently stick to the recommended limits, the risks of CHD may be lower than in non-drinkers. But 'binge' drinking is especially dangerous. This risk factor is of growing importance (page 36).

■ **Physical inactivity, obesity and generally being overweight**. A sedentary lifestyle increases the risk of CHD, but physical activity, even only moderate activity, helps prevent heart and blood vessel disorders. Regular, vigorous physical activity is especially beneficial, not least because it helps prevent obesity.

We need to be able to quantify the conditions of the body we believe are 'underweight', 'normal', 'overweight' and 'obese' for different people, if individuals are to be able to satisfactorily regulate their body weight.

How can this quantification be done?

To accurately and consistently quantify body weight in relation to health, the **Body Mass Index (BMI)** has been devised. We calculate our BMI according to the following formula:

$$\frac{\text{body mass in kg}}{(\text{height in m})^2}$$

Using our calculated BMI and Table 1.7 below, we can determine our 'weight status'.

Table 1.7 The 'boundaries' between being underweight, normal and overweight.

BMI	Status
below 18.5	underweight
18.5–24.9	normal
25.0–29.8	overweight
30.0 and over	obese

Another factor here is body shape. In fact, a basic calculation of the ratio between waist and hip size has recently been suggested as a more accurate indicator of the risk of having a heart attack in adult men and women than BMI. Those with waist to hip ratios of 0.9–1.0 or less in males and of 0.85–0.9 or less in females have a significantly lower risk of a cardiac infarction (heart attack). This correlation is attributed to abdominal fat cells being a major source of LDLs, and a source of metabolites that damage the insulin production system in the body. Note that the condition of clinical obesity is defined as having a BMI of 30 and over. The incidence of obesity has substantially increased over the past 20 years. Studies by the World Health Organization, published in 2003, estimate that 300 million are clinically obese worldwide. In the population of a developed country like the UK, 13–17% of men and 16–19% of women are clinically obese.

■ **Diabetes** is a disease that carries a significantly raised risk of the patient developing CVD, and people with this condition require especially close monitoring of their blood pressure and blood glucose to ensure they are continuously controlled within safe parameters.

■ **Other dietary issues** include the question of salt consumption. Sodium chloride is a widely used food preservative, traditionally popular in many cultures, and used in diverse prepared

Activity 1.11: Wider
reading – 'Obesity: size
matters!'

16 From the information given above, suggest whether the associations between the incidence of heart disease and diets
a high in salt
b low in antioxidants
are cases where one is shown to *cause* the other (causations), or merely where a connection is suggested (correlations).

food products. Today it plays a major part in the prevention of food poisoning when used as a preservative in packaged convenience foods and 'ready meals'. While it has this and other advantages, it does cause raised blood pressure after it has been absorbed into the blood stream and prior to its excretion by the kidneys. The Food Standards Agency recommends a salt intake of no more than 6 g per day, but it is common for an individual's intake to be twice that.

Many important vitamins have antioxidant properties. This means they supply hydrogen atoms in our body cells and so 'stabilise' unstable **radicals** – also referred to as free radicals – which are commonly occurring by-products of metabolism. An example is an oxygen atom with an unpaired electron (a superoxide radical, represented as O_{2-} or $O_2\cdot$):

$$O + e^- \rightarrow O_{2-} (O_2\cdot)$$
superoxide

- formed by the transfer of a single electron
- results in a highly destructive compound

Unstabilised radicals are highly reactive entities, and will rapidly damage important cell components. In people with diets persistently low in antioxidants, there is evidence of higher incidences of diseases such as various cancers and heart disease. On the other hand, diets that supply vitamins C and E and beta-carotene, which are efficient antioxidants, are regarded as particularly beneficial. This is the health basis of the recommendation that we eat five portions of fruit and vegetables each day.

Treatments to prevent CVD – benefits and risks

Antihypertensive medications

These include the prescription of:

- **diuretics** – substances that reduce blood pressure by decreasing blood volume. They enhance the diuretic activity of the kidneys, increasing the volume of water that is to be eliminated from the body. Patients on daily diuretics have an annual blood test to check that kidney function is not disturbed; there is a possibility that with the enhanced loss of water, essential ions in the plasma are also reduced.
- **ACE inhibitors** – substances taken daily to reduce blood pressure by enhancing **vasodilation**. When arteries are more dilated, the volume of the circulatory system is increased and so the pressure is lowered.
- **β blockers** – substances taken daily to reduce blood pressure by reducing heart rate. The heart is caused to beat less frequently.
- **calcium-channel blockers** – substances that work by widening blood vessels, so allowing more blood to flow at reduced pressure.

Patients on antihypertensive medications attend a clinic for regular measurements of blood pressure (Figure 1.21, page 21) and at least annual routine blood tests, with perhaps a check to ensure that reductions in body mass that have been requested to complement the treatments have been achieved and maintained. An additional risk of using these drugs, for a few patients, is the simultaneous triggering of change to other body systems, leading to potentially unacceptable side-effects.

Drugs to reduce blood cholesterol levels

Where blood cholesterol levels are found to be persistently at dangerous levels (serum cholesterol in excess of $5 \, mmol \, l^{-1}$), drugs such as **statins** are prescribed.

In 1970, a scientist working for a Tokyo pharmaceutical firm, who had been searching systematically for natural fungal products capable of changing the rate of particular chemical reactions within our bodies, discovered a substance that lowered the level of blood cholesterol.

17 a Explain why blood pressure is essential.

b What do people usually mean if they say they have 'blood pressure'?

Activity 1.12: Heart transplant simulation

A* Extension 1.6: Computerised tomography or CAT scan

The new fungal product turned out to be a specific inhibitor of an enzyme central to the pathway of cholesterol synthesis in the liver. This molecule was a forerunner of a series of similar compounds called statins. Statins appear to have revolutionised the prevention of heart disease where excess blood cholesterol was the cause. However, some forms of this drug are currently under suspicion of causing undesirable side-effects.

Anticoagulants and platelet-inhibitory drugs

These may be prescribed to prevent the formation of blood clots within the intact blood circulation that would otherwise lead to strokes and myocardial infarctions, where there is a strong tendency for these to occur. A daily, low-dose aspirin tablet (75 mg) may be all that is prescribed, typically after a minor heart attack or stroke. Aspirin is an anti-inflammatory substance. Other, more powerful anti-clotting drugs have been discovered, and may be prescribed along with aspirin. The obvious risks with these treatments are outweighed by the advantages, where there is a major tendency for emboli to circulate. Aspirin may carry slight risk of triggering stomach ulcers at any time.

Treatments to re-establish blood flow in damaged arteries

Coronary arteries that have been seriously damaged may be by-passed (Figure 1.26, page 25).

Coronary angioplasty (balloon angioplasty) is an alternative technique for re-establishing flow in an occluded coronary artery. Here the balloon catheter is inserted into a main artery in the patient's arm or leg, and is then guided all the way to the damaged coronary artery using X-ray observation. After the inflated balloon tip has re-established the flow and been collapsed and withdrawn, it may be necessary to shore up the damaged artery to subsequently maintain the flow of oxygenated blood. This is done by inserting a stainless steel spring (a stent) into the lumen of the artery (Figure 1.30).

The risks in this procedure are outweighed by the advantages since arteries treated using angioplasty very frequently close up again within six months, without the benefit of a stent.

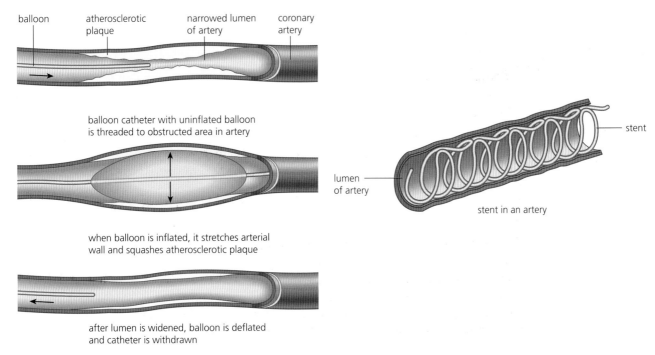

Figure 1.30 Maintaining coronary circulation by balloon angioplasty.

1.4 Lifestyles, health and risk

By **risk** we mean the chance or probability of some danger, injury or other adverse condition occurring in a given time. We have already confidently used the term in examining the factors that contribute to the medical condition of CVD and some of the lifestyle changes needed to help us to avoid it. Now we need to apply it to quantitative data, too.

First of all there is the concept of **probability**. Probability is the quantification of the chance that a stated outcome of an event will take place. Let us start with the example of a tossed coin: each time it is thrown, we recognise there is a 50% chance of it landing heads up or tails up as a result of any one throw.

By convention, probability lies on the scale between 0 (an impossibility) and 1 (a certainty) – so the coin-tossing event has a probability of 0.50 that it will land heads up. However, we commonly express probability values as percentages, as we did in this example.

So the probability P of the outcome of an event is:

$$P = \frac{\text{number of nominated outcomes}}{\text{total number of possible outcomes}}$$

18 The standard dice has six faces, only one of which has six dots. What is the probability of the dice landing 'six' face up in a single throw? Show your working.

In our coin-throwing example, let us say that there is one nominated outcome – the coin landing heads up. The total number of outcomes is the number of possible alternatives (two: heads up or tails up).

So, we can see that:

$$P = \frac{1}{2} = 0.50 = 50\%.$$

National data and individual risks

Now we can turn to health risks and examine the probabilities of contracting a fatal condition prematurely. We can use the data for the year 2004 given in Figure 1.27 (page 26). In the UK population in that year, what were the probabilities of an adult male and an adult female dying prematurely from CVD (CHD, stroke, or other)?

Look at the pie charts in Figure 1.27 now.

For males and females, the numbers dying in this way were 33% and 24%, respectively; as noted earlier, males carry a greater risk of CVD than females.

You may have noted that, like CVD, cancer is not one but many diseases – some very different.

Were the risks of premature death from a cancer in 2004 greater or smaller than from CVD in males and females?

19 The most recent figures for annual death rates can be obtained using the National Office for Statistics tables on Causes of Death. For example, for the year 2005, this table shows death rates due to CVD and other causes, among all deaths (not just in people aged 75 or younger).

	Death rates per million of the population	
	all causes	due to CVD
Males	7337	2597
Females	5188	1643

What was the risk of dying from CVD in 2005?

Epidemiology is the study of disease patterns in populations. CHD among the cardiovascular diseases has been most intensively studied because it is such a major cause of death in developed countries. For example, these studies have been responsible for exposing the risk factors we have already discussed (pages 26–30).

Further studies of CHD have shown that, among any population, the probability of premature death from heart disease (or any other CVD) is not equally distributed. Some of us, by dint of lifestyle (or sex, or genetic inheritance) for example, are more vulnerable than others. A local population study that confirms this obvious point is illustrated in Figure 1.31, and demonstrates how an individual's risk of CHD increases with the number of risk factors carried by that individual. The study was concerned with the risk factors of smoking, severe hypertension and very high blood cholesterol.

Figure 1.31 How the number of risk factors increases the risk of CHD.

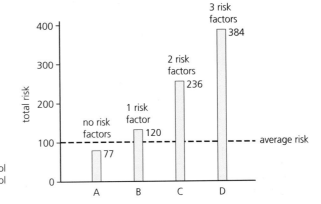

Key

A	male non-smokers,	normal blood pressure,	normal blood cholesterol
B	male smokers,	normal blood pressure,	normal blood cholesterol
C	male smokers,	normal blood pressure,	very high blood cholesterol
D	male smokers,	very high blood pressure,	very high blood cholesterol

Investigating risk

Successful correlation studies need very careful design. Firstly, a large sample must be selected to create sufficient data. Only in this way can the experimenter be certain that the results are not the products of chance. We say that the sample size must be large enough to yield statistically significant results and so be reliable.

Secondly, the sample has to be entirely representative of the population under investigation. So, for example, if some aspect of the incidence or causes of CVD is being studied, then the sample selected must reflect the community at large in terms of the balance of males and females, the proportions of different age groups, and its socio-economic composition. If any such aspect of the sample was unrepresentative of the population, conclusions may be misleading.

Clearly, the selection of samples in which to look for correlations between possible risk factors and the occurrence of disease is a major undertaking in its own right. Sampling is a technical process.

However, even when correctly set up, all correlation studies provide only *circumstantial* evidence for the role of a particular factor. The connection is not *proved* by correlation, it is merely suggested. For example, in Figure 1.32, an association is shown between the number of CHD deaths and the percentage of saturated fat in the diet, but no causal link is proved.

In fact, a risk factor is a statistical concept. After proper analysis, the experimenter can have confidence in the potential significance of the results, up to a certain level. But other types of study are required in order to establish how particular circumstances operate unfavourably on the functioning body to *cause* a diseased state, if in fact they do. Typically, confirmations are obtained from:

20 Explain what is meant by 'control group' in a medical investigation.

■ **clinical trials**, which investigate, among other things, the biochemical and/or structural changes that are induced in patients
■ **intervention studies**, which work with vulnerable groups of patients, changing their habits and lifestyles, and establishing the extent to which the incidence of disease can be reduced compared with rates in other patients. These others are, in effect, a control group.

Activity 1.13: Analysis of changing mortality in the twentieth century

The outcomes of such studies and investigations into CHD are that individuals increasingly use scientific knowledge about the effects of diet (including obesity indicators, page 29), exercise and smoking to reduce their risk of CHD, and with total confidence.

Figure 1.32 Correlation studies in the epidemiology of CHD.

Correlation study concerning saturated fat intake and the incidence of CHD

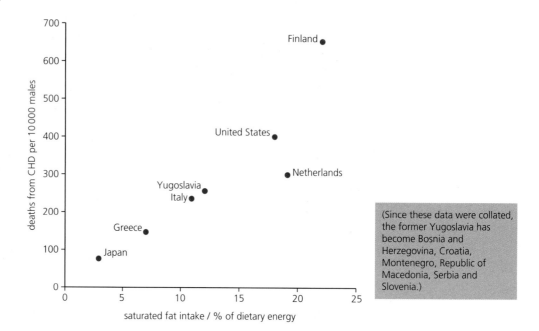

(Since these data were collated, the former Yugoslavia has become Bosnia and Herzegovina, Croatia, Montenegro, Republic of Macedonia, Serbia and Slovenia.)

Understanding correlations – the differences between a positive correlation (left) and a negative correlation (right)

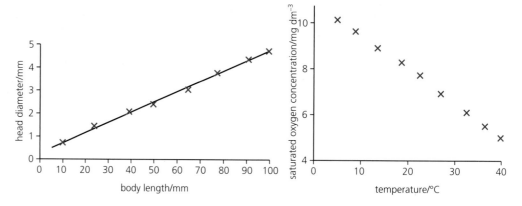

Another complication – perceptions about risks

From time to time, a tragic accident occurs and receives huge publicity. For example, in 2000 there was a major accident on the UK railway system at Hatfield in Hertfordshire, just north of London (Figure 1.33). A train carrying about 200 passengers, and travelling at a speed of 115 mph, was derailed and many of the carriages tipped over. Horrendous destruction was caused to the train carriages and the surrounding track. Four people were killed – they were thought to have been travelling in the buffet car which had its roof ripped off – and there were 30 'walking wounded', who were taken to hospital. There, a spokesman said later that the major casualties had suffered 'lumps, bumps and fractures'.

Figure 1.33 The Hatfield rail disaster, 2000.

An investigation by the Transport Police was immediately launched. It was said that the work of terrorists or vandals could not be ruled out. National and international terrorism was (and remains) a major concern – the risk of such disruption and destruction was a serious worry for many people. Quite quickly, a cracked rail was discovered and established as the cause of the accident, which led to a huge programme of rail checks and replacements. Speed limits were imposed across the whole rail network to allow this, causing all timetables to be changed and the public to be seriously inconvenienced for a significant period.

One inevitable outcome was that many more people had to (or chose to) use the road system for many weeks, to get about. Look at the comparative figures for passenger deaths associated with various forms of transport (Table 1.8). On any day in the year, typically eight or nine people are killed on the roads in the UK. Following high-profile accidents such as that at Hatfield, many people feel far safer inside their car than on a train, but in fact rail travel is almost as safe as air travel. Actual risks are often overestimated, as in this case of rail travel dangers.

Table 1.8 Summary of passenger transport risks (in the European Union, 2001–02).

	Deaths per 100 million person travel hours
Road	28
Ferry	16
Rail	8
Air (civil aviation)	2

Data from *Transport safety performance in the EU – a statistical overview*, by the European Transport Safety Council's Transport Accidents Statistics Working Party.

On the other hand, major risks can also be underestimated or ignored. For example, despite the attention now given to reducing CVD by change in lifestyle and by regular medical checks, followed up with appropriate preventative measures (treatments for hypertension, medication to reduce blood cholesterol, and so on), many people are also increasing their consumption of alcohol. The change in drinking habits that has occurred with increasing affluence (for many) in countries such as the UK, the ease of buying wines and beers for consumption at home, and the relative decline in the costs of alcoholic drinks, have all contributed to a developing, major public health problem. Yet many underestimate the risks involved (Figure 1.34).

Check back (page 29) to remind yourself of the consequences for CVD of high and continuing alcohol abuse.

Activity 1.14: Research a case of risk – prions as disease agents

Figure 1.34 Raising alcohol consumption – the facts.

THE TIMES

Tuesday June 5 2007 timesonline.co.uk No 69030

Crackdown on middle class wine drinkers

Higher alcohol taxes are needed, say doctors

New campaign to combat drunkenness

Middle-class wine drinkers will be the focus of government plans to make drunkenness as socially unacceptable as smoking, *The Times* has learnt.

Under the plans published today a fresh audit is to be conducted by the Government into the overall costs of alcohol abuse to society and the National Health Service.

The assault on Middle England's drinking habits is part of a three-strand approach, which will also target under-age drinking and heavy alcohol consumption among those aged 18–24.

"We want to target the older drinkers, those that are maybe drinking one or two bottles of wine at home each evening. They do not realise the damage they are doing to their health and that they risk liver disease," a Whitehall source said.

"There are growing numbers of people turning up in hospital with drink-related diseases and drink-related injuries. They are getting younger and more of them are turning up needing treatment," the source added.

Richard Ford, David Rose, Patrick Foster.

Getting the diet right for health

We describe the food we eat as our diet. A balanced diet consists of the essential nutrients in the correct proportions. A diet that is deficient in one or more essential nutrients, or provides the wrong balance of nutrients, causes malnutrition. Incidentally, malnutrition is experienced by very many human communities around the world (as you are already aware), but malnutrition is a misfortune that can befall any organism in nature – and frequently does.

Food should provide an appropriate amount of the following six components on a regular basis.

1 Metabolic fuel

Metabolic fuel is normally supplied from carbohydrates and fats (lipids), but it can also be obtained from proteins. Chemical energy in all of these fuels is transferred by respiration, but proteins must first be deaminated. This means that combined nitrogen, in the form of the amino group ($-NH_2$), is removed first, before respiration occurs.

The energy value of foods is expressed in **joules** (J) or kilojoules (kJ) ($1\,kJ = 1000\,J$). The joule is defined in terms of heat energy; $4.18\,J$ of heat energy are required to raise 1 g of water through 1 °C. The energy typically provided by individual nutrients is estimated experimentally, using the bomb calorimeter. The amounts of energy that major food substances might yield are shown in Table 1.9.

Table 1.9 Energy yields of major food groups.

Food	Energy/kJ g^{-1}
carbohydrates, e.g. glucose or sucrose	16
fats	37
proteins	17

An earlier term for the amount of energy in food substances was the calorie. This term may often be read on commercial packs of food ($1\,calorie = 4.18\,J$).

The body's requirements for energy are determined by four factors:

■ the background rate of metabolism, known as **basal metabolic rate** – this is the amount of energy required to maintain brain function and the rest of the body's basic metabolism 'ticking over' comfortably, at normal temperatures, when at rest; it is the rate against which we compare the energy demands of various activities

- how physically active we are in work and leisure – additional energy is needed for physical activity
- our age and the amount of growth and repair work the body has to support – the rapid growth that occurs in adolescence makes an additional demand of about 8 kJ per hour
- our sex – usually, females require less energy than males, but in pregnancy and during lactation (breast feeding of an infant) female requirements are greatly increased.

Age range/ years	Estimated average requirements for energy/kJ	Protein /g	Calcium /mg	Iron/mg	Zinc/mg	Vitamin A/μg	Thiamin/ mg	Vitamin B6/mg	Folic acid /μg	Vitamin C/mg
Males										
15–18	11 510	55.2	1000	11.3	9.5	700	1.1	1.5	200	40
19–50	10 600	55.5	700	8.7	9.5	700	1.0	1.4	200	40
Females										
15–18	8830	45.0	800	14.8	7.0	600	0.8	1.2	200	40
19–50	8100	45.0	700	14.8	7.0	600	0.8	1.2	200	40
pregnant	+800	+6.0				+100	+0.1		+100	+10
lactating	+2200	+11.0	+5500		+6.0	+350	+0.2		+60	+30

Table 1.10 Recommended daily intakes.

Food outlet type	Food item	Approximate energy content/kJ
burger restaurant	hamburger	1220
	quarter pounder	1760
	cheeseburger	1260
	double burger with cheese	2270
	chicken in a bun	1810
	french fries	880
	grill	2180
	apple pie	1010
	hotcakes with butter and syrup	2100
	vanilla shake	1470
fried chicken restaurant	chicken (two pieces, with fries)	3280
	spare ribs	1600
	jacket potato	630
pizza restaurant	cheese and tomato traditional pizza	2940
	deep pan pizza	5250
fish and chip shop	cod fried in batter, with chips	3150
	can of coke or lemonade	181
	can of diet coke or lemonade	1.6
	chocolate bar	1349
	packet of crisps	564

Table 1.11 Food values of selected fast-food items.

In Table 1.10 the recommended requirements for energy and certain nutrients are compared for some categories of humans. Eating habits in developed countries are currently such that a significant part of our diet may be taken in the form of snacks and convenience foods, sometimes from fast-food outlets. Table 1.11 gives the food value of some familiar snacks and fast-food items.

> **21** Using the data in Tables 1.10 and 1.11, calculate the total number of kilojoules that you currently acquire via your eating of snacks and convenience foods taken between meals on a typical day, and express this as a percentage of your daily requirement for energy.

2 Combined nitrogen

Combined nitrogen is required for the building of proteins, and is taken in as proteins or amino acids in the diet. Dietary protein is the chief source of the 20 amino acids from which a mammal's own proteins are built up.

3 Water

Since 70–90% of the body is water, it must make up a significant part of the diet.

4 Roughage or dietary fibre

Dietary fibre is mostly of plant origin, consisting of cellulose from cell walls. It cannot be digested; its value is as bulk that stimulates food movement through the alimentary canal. Additionally, a high-fibre diet obtained from fruit and vegetables is positively correlated with a lowered risk of gut cancers and of coronary heart disease.

5 Essential minerals

These include major minerals like calcium, iron and phosphate ions, which are needed for the construction of body tissues or are combined in metabolites essential for many metabolic processes. Minerals also include the micronutrients such as manganese, which are often co-factors in the functioning of particular enzymes.

6 Vitamins

Vitamins are organic compounds that are required in only tiny amounts. Most function as coenzymes in the body. For example, vitamin C is essential for collagen fibre synthesis and healthy connective tissue. Vitamins cannot be manufactured in the body, so their absence from the diet tends to have marked effects, known as deficiency diseases. The deficiency symptoms due to lack of vitamin C are scurvy, bleeding from gums, and poor wound healing. Vitamin C is obtained from vegetables (potatoes and green vegetables) and fruits, particularly citrus fruits. It is possible to estimate the vitamin C content of foods and drinks (see HSW 1.3).

The vitamin content of foods

HSW 1.3: Criteria 2, 3, 4, 5 and 6 – Planning and conducting an investigation of the vitamin C content of fruit and vegetables

Vitamin C is a relatively small organic molecule of molecular formula $C_6H_8O_6$. The molecule is known as ascorbic acid, but it does not contain a free carboxyl group (–COOH) as true organic acids do. However, it is 'sharp' to the taste and does form salts. It is a good reducing agent. Ascorbic acid is water soluble, and it is destroyed by cooking, particularly in alkaline conditions. Humans are among the few animals unable to form their own vitamin C (as are the other higher primates and guinea pigs), and so require it in foods, but it is not widely distributed in food items.

Energy intakes and the conditions of being overweight or underweight

Table 1.12 shows that the daily estimated average requirement for energy of people in the age range 15–50 years varies with their sex.

Table 1.12 Average daily energy requirements for men and women.

Age range/years	Estimated average daily requirement for energy/kJ	
	men	women
15–18	11 510	8830
19–50	10 600	8100

These are estimated average requirements – whether we as individuals maintain a good weight for our height (that is, a BMI of 20–24.9, page 29), or instead become overweight or underweight, is more likely to depend on the *balance* between our energy intake and energy use than on the amount of energy taken in per se (Table 1.13).

Table 1.13 Maintaining a healthy body weight is achieved by balancing energy intake with energy use.

Body mass condition	Energy intake	Energy use
underweight	inadequate, possibly due to illness, poor diet, or to an eating disorder	excessive, due to physical activity, stressful existence, or to a high basal metabolic rate
normal	balanced	
overweight	excessive, probably due to over-eating and drinking	inadequate, probably due to a lack of physical activity

Being overweight (and possibly obese) is most prevalent in people of developed countries. Here is a recognised public health problem for all age groups, as we have already noted. Overweight people have increased risk of coronary heart disease, gall bladder disease, high blood pressure and diabetes. Most people who become overweight do so due to persistent consumption of too many energy-rich items of diet in relation to their energy requirements.

Being underweight, among people in the developed world, may be due to an eating disorder. The conditions known as anorexia nervosa and bulimia nervosa are thought to be on the increase, particularly among young Caucasian females from middle or upper social classes. In anorexia, deliberate dieting, and sometimes deliberate vomiting, lead to serious weight loss and even the loss of consecutive menstrual cycles. Patients have an obsessive fear of gaining weight or becoming fat; they see themselves as much fatter than they actually are.

Figure 1.35

Girls gripped by worry over looks

Many teenage girls worry about their appearance more than anything else in their lives. Weight and appearance are by far the main concerns for girls, eclipsing family problems, difficulties with friends, health, careers and school, according to an **Exeter University survey**.

'*The overwhelming majority of those who say they wish to lose weight have no medical weight problem at all, and some are underweight,*' said researcher John Balding.

Researchers from the Schools Health Education unit questioned 37 500 young people aged 12–15 years, and concluded that girls were much more worried about their appearance than boys.

The study concluded that, to help counter the problem, a greater diversity of body shapes among actresses and models needed to be seen on television and in newspapers.

by journalist Helen Carter
First published in *The Guardian*, 16/11/98

Activity 1.15:
Estimating your kilojoule
intake in snacks

HSW 1.4: Criterion 10 –
Considering ethical
issues in the treatment
of humans

22 Explain why failures
in the maintenance
of body collagen
may lead to
haemorrhages.

In bulimia, periods of excessive eating ('binge eating') are followed by self-induced vomiting and use of laxatives to achieve weight control. Here, patients do not necessarily lose excessive weight and their menstrual cycles remain normal.

These two conditions are believed to have more to do with anxiety about maturation and sexuality than with diet, in many cases (Figure 1.35).

■ Extension: A note on dietary reference values

The UK Government provides dietary guidelines for nutrients in the *Manual of Nutrition*, published by The Stationery Office and obtainable from its Publications Centre. It is a reference book well worth consulting. For example, you will discover there that the term 'recommended daily amounts' has been replaced by 'dietary reference values'.

The term dietary reference value (DRV) has been introduced to assess the adequacy of diets of different groups of people. DRVs are a product of three values:

- **estimated average requirements (EAR)**, an estimate of the average need for food energy or a nutrient
- **reference nutrient intake (RNI)**, the amount of a nutrient that any individual would need, even those with high needs
- **lower reference nutrient intake (LRNI)**, the amount of a nutrient sufficient for individuals with low needs.

Tables of the values of EARs and RNIs are given in the booklet *Manual of Nutrition*, where their involvement in the assessment of diets of groups of people (such as infants, adolescents and older people) is discussed. They provide a range of values within which a healthy balanced diet should fall.

Carbohydrates and fats as our chief energy source

We have listed above the components of a balanced diet – the substances that provide us with essentials. Some of these components are also described as **essential nutrients** – but not all of them. Essential nutrients are the components of a diet that must be present because they cannot be entirely or adequately produced by the body from other substances taken in as food, in sufficient quantities for survival and good health.

Remember, the other role of nutrients is to provide the materials needed to build up cells and their components. This is additional to the energy that some are able to transfer.

Accordingly, the **essential nutrients in our diet** are:

- amino acids, obtained from proteins
- fatty acids, obtained from lipids
- minerals
- vitamins
- water.

It may surprise you to notice that carbohydrates are omitted from this list. This is because the energy they provide can equally be obtained from fatty acids and from proteins.

However, although carbohydrate is not an essential, most diets not only include it, but in fact many people gain the bulk of their essential metabolic energy from it. Carbohydrates are an extremely common source of metabolic fuel available to an organism – very often the chief source, in fact. Table 1.14 shows the chief metabolic fuels in diets from around the world.

Table 1.14 Main dietary sources of energy around the world.

Source	Locality/region	Environmental features/food value
rice, *Oryza sativa*	grown extensively throughout Asia, e.g. China (35% of world production)	■ cultivated in China for 5000+ years ■ grown in paddy fields of standing water ■ grains provide 30%+ starch and about 2.5% protein
wheat, *Triticum aestivum* (bread wheat)	grown widely, but other varieties common too	■ requires moderate moisture and cool weather for early growth; then sunny, dry months ■ provides flour of 70% starch and 12% protein
cassava, *Manihot esculenta*	grown under 'slash and burn' cultivation in tropical African countries, e.g. Nigeria, Congo and Rwanda	■ survives in poor soil with little attention ■ provides foliage for livestock (70% protein), and swollen roots of 30% starch (no protein or lipid); these are boiled and then pounded into a thick paste
maize, *Zea mays*	originated in South America (modern Mexico), now cultivated widely to be a mainstay food in many communities	■ modern varieties are hybrid seeds, requiring advanced agricultural practices – modern cultivation, fertilisers and pesticides, and careful harvesting ■ most varieties yield starch and oil, but some yield protein also

In the UK, dietary energy intake comes from both carbohydrates and lipids. In fact, the dietary guidelines for daily energy state the percentages that should be provided from carbohydrates and fats (Table 1.15).

Table 1.15 Dietary reference values for fats and carbohydrates as a percentage of daily energy intake.

	Percentage of daily energy intake/%
Fat	35
saturated fats	11
unsaturated fats	21.5
others	2.5
Carbohydrate	50
sugars	13
starch	37

Activity 1.16: Dieting and essential nutrients

Fats are specified because a wide range of different fatty acids are used in the cells, tissues and organs in the biosynthesis of metabolic intermediates, membrane lipids including very many receptors, and many hormones. The sex hormones are of lipid origin, as are others that are critically involved in the regulation of metabolism and the operation of the body's physiology. Further, the key roles of unsaturated fatty acids in a healthy blood circulatory system have already been highlighted. That is, monounsaturated fats help to remove circulating LDLs ('bad cholesterol'), and polyunsaturated fats further increase the efficiency of this removal.

Omega-3 fatty acids – a case of conflicting evidence

Omega-3 fatty acids are a select group of naturally occurring polyunsaturated fatty acids. They are chemically special in that they have between three and six double bonds in the hydrocarbon tail, and, in particular, the first double bond is *always* positioned between the third and the fourth carbon atom from the opposite (omega) end of the hydrocarbon chain to the carboxyl group, as shown in Figure 1.36. A number of omega-3 fatty acids occur in plant and fish oils, and are thought to be particularly beneficial to health.

Some dieticians argue that omega-3 polyunsaturated fatty acids, while not influencing the levels of cholesterol circulating in the blood, appear to help prevent heart disease by reducing

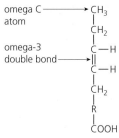

Figure 1.36 An omega-3 fatty acid.

HSW 1.5: Criterion 6 – Omega-3 fatty acids and heart disease: another fishy story?

A* Extension 1.7: Obesity: An overblown epidemic? A study of conflicting evidence

the tendency of the blood to form clots, and by giving us healthy, well-functioning plasma membranes around cardiac muscle fibres. This claim is based upon studies (known as **cohort studies**) of people who have a high daily intake of these fatty acids, such as Greenland Inuit peoples.

A cohort is a group of people who share a common characteristic or experience within a defined time period. In the case of the Inuit people, the common experience is a diet exceptionally rich in oily fish meat. Associated with this diet (and active lifestyle) is a very much reduced likelihood of an acute myocardial infarction (heart attack), compared with other human groups.

It is estimated by dieticians who favour such cohort studies that we need about 0.2 g per day (1.5 g per week) of omega-3 fatty acids, and this need will be met if we eat sufficient oily fish regularly. Walnuts are a plant source rich in omega-3 fatty acids.

However, other investigations have employed **randomised control trials (RCT).** RCTs are a commonly used method of evaluating the effectiveness of medicines and medical procedures. Here, treatments are allocated to subjects at random, and both the patients and those who conduct the trials remain unaware as to which individuals receive the experimental treatment and which receive a placebo (an alternative substance with no therapeutic effect). Both groups are treated identically, except on the issue of which substance is administered.

The outcomes of RCTs suggest that it is difficult to show the clear benefits of omega-3 fatty acids claimed from the cohort studies. Since the evidence is conflicting, this issue may remain a controversial one for some time. Note that the issue of statistically significant drug testing procedures is further discussed in Chapter 4.

1 End-of-topic test

(full End-of-topic tests are available on the DL Student website)

1 Read through the following account and then write in the spaces provided the most appropriate word or words to complete each statement. (8)

The water molecule is composed of one atom of oxygen and two atoms of hydrogen, combined by _____ electrons (_____ bonding). However, the water molecule is _____ rather than linear, and the nucleus of the oxygen atom draws _____ (negatively charged) away from the hydrogen nuclei (positively charged) – with an interesting consequence. Although overall the water molecule is electrically neutral, there is a net _____ charge on the oxygen atom and a net _____ charge on the hydrogen atoms. In other words, the water molecule carries an _____ distribution of electrical charge within it. This arrangement is known as a _____ molecule.

2 The diagram below shows the structure of two monosaccharide sugars, glucose and fructose.

glucose

fructose

a Draw a diagram to show the structure of the disaccharide formed when these two molecules bond. (2)

b What type of reaction has occurred, and what type of bond has been formed? (2)

3 The table below refers to three polymers. In each box, place a tick to show if the statement is correct for that polymer. Place a cross to indicate if the statement is incorrect. (6)

Statement	Glycogen	Amylose	Amylopectin
polymer of glucose monomers			
found in cells of plants			
built from unbranched chains only			
formed by condensation reaction			
energy store found in muscles			
glycosidic linkages present			

4 Explain in your own words the difference between a saturated and an unsaturated lipid. (3)

5 The graph below represents the changing blood pressure throughout the circulation, as it is often shown in medical sources. Note that pressure falls as the blood flows through the arteries, arterioles, capillaries and veins, and that pressure is recorded in 'millimetres of mercury' (mmHg).

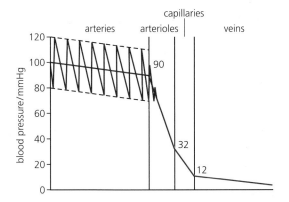

a The kilopascal (kPa) is generally used by scientists to measure pressure. Given that 1 mmHg = 0.13 kPa, express in kilopascals:
  i the highest pressure recorded in the arteries
  ii the fall in pressure that occurs whilst blood flows through the capillaries.
b Explain fully the reason for the changes in blood pressure as blood flows through:
  i the arteries
  ii the capillaries.
c Explain how backflow of blood in the veins is prevented, despite the low pressures there and the external pressures imposed by surrounding body tissues such as the contracting muscles. (6)

6 In the following account certain words have been omitted. Read the text and then fill in the blanks. (6)

The molecules that make up a substance are constantly on the move in a _____ way. It is these movements that cause molecules to be dispersed from areas of _____ concentration to areas of _____ concentration, a process called diffusion. We say that molecules diffuse along a concentration gradient. Diffusion occurs freely across thin, _____ membranes, too. Each type of molecule diffuses along its concentration gradient; diffusion in one direction is _____ of diffusion of other molecules in an opposite direction. An example of this process is the inward diffusion of _____ into a respiring cell and outward diffusion of _____. Another molecule constantly on the move across cell membranes is water. Movement of water is by _____, a special case of diffusion, due to the presence of a _____ -permeable membrane. Free water molecules

diffuse across such a membrane unhindered. Dissolved substances attract a group of the _____ molecules around them, held by _____ bonds, and their free movements are _____.

7 In an experiment to investigate whether a particular concentration of caffeine will increase the heart rate of a water flea such as *Daphnia*, describe what control you would require as part of your procedure, and why. (6)

8 The blood-clotting mechanism may be represented by a simple flow diagram.

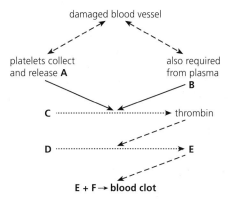

a From the list below, select substances that are the essential components of the blood-clotting process identified in the flow diagram by the letters **A → F**.
cholesterol; prothrombin; glucose; fibrin fibres; adrenalin; calcium ions + vitamin K; blood cells; thromboplastin; amino acids; prothrobin; insulin; fibrinogen; vitamin C
b Suggest an advantage to the body of the complexity shown by the clotting mechanism. (8)

9 Studies of the incidence of conditions such as cardiovascular disease include investigations of the risks of contracting the condition in different populations.

a In the design of statistically significant studies, explain why the following are given special attention:
  i size of samples
  ii composition of samples.

b Given the establishment of a statistically significant association between a particular factor and the incidence of a disease, outline two types of study that are necessary to confirm the existence of a causal relationship. (8)

2 Cells, genes and health

STARTING POINTS

■ The cell is the basic unit of life. Too small to be seen with the naked eye, cells are viewed using a compound or electron microscope.
■ The contents of cells are contained by a membrane, known as the plasma membrane. The membrane forms a barrier across which all substances entering and leaving the cell must pass.
■ The organic compounds that make up living things include the proteins and nucleic acids. Many proteins are enzymes. Enzymes are biological catalysts that make possible chemical changes under the conditions in cells.
■ The nucleus of a cell houses the chromosomes, which contain the hereditary material, a nucleic acid known as DNA. DNA carries the coded instructions for the cell.
■ Genetics is the study of the inheritance of characteristics by offspring from their parents.

■ 2.1 The cell membrane – structure and function

1 Identify the essential processes characteristic of living things.

The cell is the basic unit of living matter – the smallest part of an organism that we can say is alive. It is cells that carry out the essential processes of life. We think of them as self-contained units of structure and function. Many organisms are made of a single cell, including the bacteria, while others are multicellular in structure, including the mammals. Much of the biology in this book is about multicellular organisms, such as humans, and the processes that go on in them. But remember, single-celled organisms carry out all the essential functions of life too, occuring within the confines of a single cell.

You will have observed living plant and animal cells, using a compound light microscope, and will already appreciate that they are extremely small structures. Appropriate units are required if we are to measure them (Table 2.1).

The metre (m) is the standard unit of length used in science (it is an internationally agreed unit, or SI unit). We are familiar with the subdivision of the metre into millimetres (mm) – units of length that are one thousandth (10^{-3}) of a metre. However, millimetres are also too large to use in cell measurement – a further subdivision is required. In fact, the dimensions of cells are expressed in micrometres or microns (μm). One micrometre is one thousandth (10^{-3}) of a millimetre, which gives us an idea about how small cells are. Bacteria are really small, typically 0.5–10 μm in size, whereas the cells of plants and animals are often in the range 50–150 μm, or larger.

2 How many cells of 100 μm diameter fit side by side along a millimetre?

Table 2.1 Units of length used in microscopy.

1 metre (m)	= 1000 millimetres (mm)
1 mm	= 1000 micrometres (μm or microns)
1 μm	= 1000 nanometres (nm)

Observing the structure of cells

Despite the fact that cells are extremely small, by using the light microscope we can see that they consist of a nucleus surrounded by cytoplasm, contained within a membrane (Table 2.2).

Table 2.2 Introducing cells.

Structure	Description
nucleus	The **nucleus** is the structure that controls and directs the activities of the cell.
cytoplasm	The **cytoplasm** is the site of the chemical reactions of life, which we call **metabolism**.
plasma membrane	The cell membrane is known as the **plasma membrane**. This structure holds the cell contents together, and is a barrier to substances entering and leaving the cell.

Activity 2.1: Measuring the size of cells

To resolve details of cell structure, the light microscope is not competent. It is to electron microscopes, and particularly to the transmission electron microscope, that cytologists turn, in this case. We discuss this instrument in Chapter 3 (page 99).

The structure of the plasma membrane

The plasma membrane is the structure that maintains the integrity of the cell (it holds the cell's contents together). It is also the barrier across which all substances entering and leaving the cell must pass. So the membrane's properties, based on its chemical composition and structure, are all important in the operation of the cell.

The plasma membrane is made almost entirely of protein and lipid, together with a small and variable amount of carbohydrate. In Figure 2.1, a model of the molecular structure of the plasma membrane, known as the **fluid mosaic model**, is illustrated. The plasma membrane is described as a *mosaic* because the proteins are clearly scattered about in this pattern, and *fluid* because the components (lipids and proteins) are able to move past each other in a linear plane.

Figure 2.1 The fluid mosaic model of the plasma membrane.

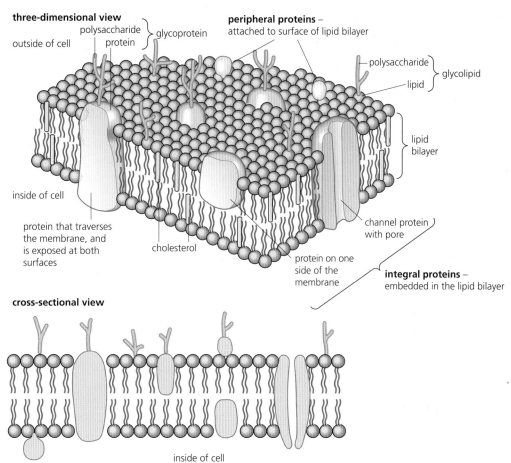

What is the evidence for this model of membrane structure?

It was in 1972 that two cytologists, S. J. Singer and G. L. Nicolson, proposed the fluid mosaic model of membrane structure. The model was built up from a body of evidence that had accumulated over a period of time, from studies of cell structure (cytology), cell biochemistry and cell behaviour (cell physiology). This evidence, in the form of ten important observations, is reviewed next.

1 Cell contents are observed to flow out when the cell surface is ruptured, as illustrated in Figure 2.2 in a damaged red blood cell. This confirms the presence of a physical barrier around the cytoplasm that is, under normal circumstances, well able to contain and protect the cell contents.

Figure 2.2 A red cell with a damaged plasma membrane.

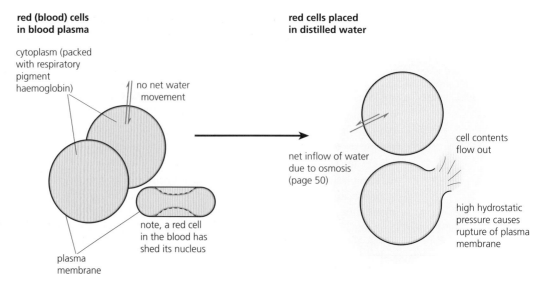

red (blood) cells in blood plasma

cytoplasm (packed with respiratory pigment haemoglobin)

no net water movement

note, a red cell in the blood has shed its nucleus

plasma membrane

red cells placed in distilled water

net inflow of water due to osmosis (page 50)

cell contents flow out

high hydrostatic pressure causes rupture of plasma membrane

2   Water-soluble compounds enter cells less readily than compounds that dissolve in lipids (these will be non-polar compounds and hydrophobic substances). This implies that lipids are a major component of the plasma membrane.

3   Lipids obtained from cell membranes consist of a type of compound known as a **phospholipid**. The chemical structure of a phospholipid is shown in Figure 2.3.

Phospholipid has a 'head' composed of a glycerol group, to which is attached one ionised phosphate group. This latter part of the molecule has **hydrophilic** (water-loving) properties. For example, **hydrogen bonds** readily form between the phosphate head and water molecules. The remainder of the phospholipid comprises two long, fatty acid residues consisting of hydrocarbon chains. These 'tails' have **hydrophobic** (water-hating) properties. So phospholipids are unusual in being partly hydrophilic and partly hydrophobic.

Figure 2.3 The chemical nature of a phospholipid.

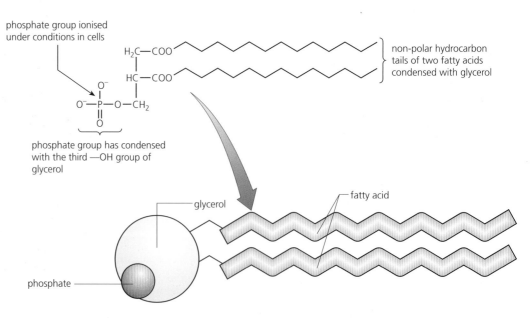

phosphate group ionised under conditions in cells

H_2C—COO

HC—COO

non-polar hydrocarbon tails of two fatty acids condensed with glycerol

O⁻—P—O—CH_2

phosphate group has condensed with the third —OH group of glycerol

glycerol

fatty acid

phosphate

4   The behaviour of phospholipids when added to water was predicted from this structure – and is demonstrated in practice. With a small quantity of phospholipid in contact with water, these molecules form a monolayer that floats with the hydrocarbon tails exposed above the water (Figure 2.4). When more phospholipid is available, the molecules arrange themselves as a bilayer, with the hydrocarbon tails facing together. This latter is the situation in the plasma membrane model. Furthermore, in the lipid bilayer, attractions between the hydrophobic

hydrocarbon tails on the inside, and between the hydrophilic glycerol/phosphate heads and the surrounding water molecules on the outside, result in a stable, strong barrier.

Figure 2.4 Phospholipid molecules and water – the formation of monolayers and bilayers.

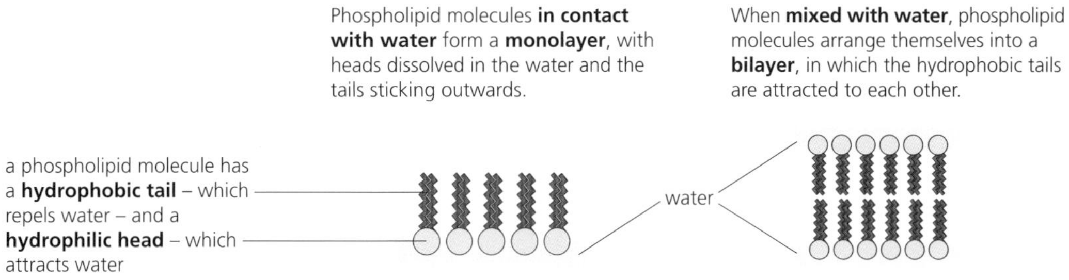

Phospholipid molecules **in contact with water** form a **monolayer**, with heads dissolved in the water and the tails sticking outwards.

When **mixed with water**, phospholipid molecules arrange themselves into a **bilayer**, in which the hydrophobic tails are attracted to each other.

a phospholipid molecule has a **hydrophobic tail** – which repels water – and a **hydrophilic head** – which attracts water

water

3 Draw a copy of the diagrammatic cross-section of the fluid mosaic membrane (part of Figure 2.1). Label the phospholipid bilayer, cholesterol, glycoprotein, integral protein and peripheral protein.

5 Chemical analysis of plasma membranes has also shown that, although a significant proportion of lipid is present, there is insufficient in total to cover the whole of the cell surface in a bilayer. Furthermore, protein is also present as a major component. The proteins of plasma membranes are globular proteins (page 63).

6 Work on the extraction of protein from plasma membranes indicated that, while some occur on the external surfaces and are easily extracted, other proteins occur buried within or across the lipid bilayer. These are difficult to extract.

7 Electron micrographs (EMs) of plasma membrane fragments, which had by chance split down the midline, showed that, some proteins occur buried within or across the lipid bilayer (Figure 2.5). Proteins that occur partially or fully buried in the lipid bilayer are described as **integral proteins**. Those that are superficially attached on either surface of the lipid bilayer are known as **peripheral proteins**. The roles of membrane proteins have also been investigated, and these are diverse. Membrane proteins may be channels for transport of metabolites, or enzymes and carriers, and some may be receptors or antigens.

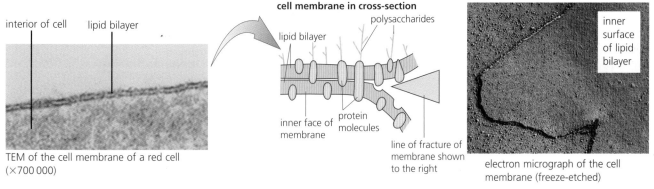

interior of cell   lipid bilayer

TEM of the cell membrane of a red cell (×700 000)

cell membrane in cross-section

lipid bilayer

polysaccharides

inner face of membrane

protein molecules

line of fracture of membrane shown to the right

inner surface of lipid bilayer

electron micrograph of the cell membrane (freeze-etched)

Figure 2.5 Plasma membrane structure; evidence from the electron microscope.

8 Experiments in which specific components of membranes are 'tagged' by reaction with marker chemicals (typically fluorescent dyes) show that the component molecules within membranes are continually on the move. The membrane's structure can truly be described as 'fluid'.

9 Lipid bilayers have been found to contain molecules of a rather unusual lipid, in addition to phospholipids. This lipid is known as **cholesterol** (page 27). Cholesterol has the effect of disturbing the close-packing of the phospholipids, thereby increasing the flexibility of the membrane.

10 On the outer surface of the cell, antenna-like carbohydrate molecules form complexes with certain of the membrane proteins (forming **glycoproteins**) and lipids (**glycolipids**). The functions of these complexes have since been shown to be cell–cell recognition, or as receptor sites for chemical signals. Others are involved in the binding of cells into tissues.

DL
www
Activity 2.2: Pouring oil on troubled waters

HSW 2.1: Criterion 1 – Use of theories, models and ideas to develop and modify scientific explanations

Movement across the plasma membrane

Movement of molecules across the plasma membranes of living cells is continuous and heavy. Into and out of cells pass water, respiratory gases (O_2 and CO_2), nutrients such as glucose, essential ions, and excretory products. Cells may secrete substances such as hormones and enzymes, and they may receive growth substances and certain hormones.

Plant cells secrete the chemicals that make up their walls through their cell membranes, and assemble and maintain the wall outside the membrane. Certain mammalian cells secrete structural proteins such as collagen, in a form that can be assembled outside the cells in the production of connective tissues, for example.

In addition, the plasma membrane is where the cell is identified by surrounding cells and organisms. For example, protein receptor sites are recognised by hormones, neurotransmitter substances from nerve cells, and other chemicals, sent from other cells. Figure 2.6 is a summary of this movement, and also identifies the possible mechanisms of transport across membranes, into which we need to look further.

Figure 2.6 Movements across the plasma membrane.

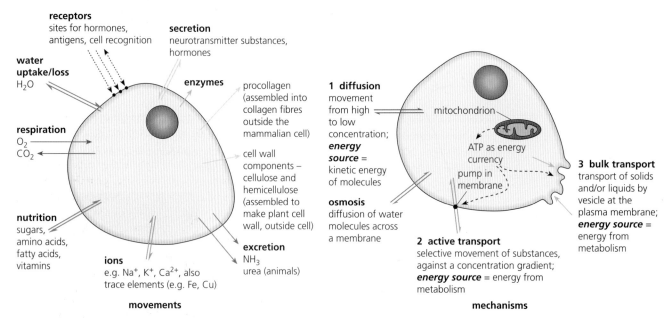

receptors
sites for hormones, antigens, cell recognition

secretion
neurotransmitter substances, hormones

water uptake/loss
H_2O

enzymes

procollagen (assembled into collagen fibres outside the mammalian cell)

respiration
O_2
CO_2

cell wall components – cellulose and hemicellulose (assembled to make plant cell wall, outside cell)

nutrition
sugars, amino acids, fatty acids, vitamins

ions
e.g. Na^+, K^+, Ca^{2+}, also trace elements (e.g. Fe, Cu)

excretion
NH_3
urea (animals)

1 diffusion
movement from high to low concentration;
energy source = kinetic energy of molecules

osmosis
diffusion of water molecules across a membrane

mitochondrion

ATP as energy currency
pump in membrane

2 active transport
selective movement of substances, against a concentration gradient;
energy source = energy from metabolism

3 bulk transport
transport of solids and/or liquids by vesicle at the plasma membrane;
energy source = energy from metabolism

movements

mechanisms

1 Movement by diffusion

The atoms, molecules and ions of liquids and gases undergo continuous random movements. Given time, these movements result in the complete mixing and even distribution of the components of a gas mixture, and of the atoms, molecules and ions in a solution. So, for example, from a solution we are able to take a tiny random sample and analyse it to find the concentration of dissolved substances in the whole solution – because any sample has the same composition as the whole. Similarly, every breath we take has the same amount of oxygen, nitrogen and carbon dioxide as the atmosphere as a whole.

> **Diffusion is the free passage of molecules (and atoms and ions) from a region of their high concentration to a region of low concentration.**

Where a difference in concentration has arisen between areas in a gas or liquid, random movements carry molecules from a region of high concentration to a region of low concentration. As a result, the particles become evenly dispersed. The energy for diffusion comes from the **kinetic energy** of molecules. 'Kinetic' means that a particle has this energy because it is in continuous *motion*.

Diffusion in a liquid can be illustrated by adding a crystal of a coloured mineral to distilled water (Figure 2.7). Even without stirring, the ions become evenly distributed throughout the water. The process takes time, especially as the solid has first to dissolve.

Figure 2.7 Diffusion in a liquid.

1 A crystal of potassium permanganate (potassium manganate(VII), $KMnO_4$) is placed in distilled water.

2 As the ions dissolve, random movements disperse them through the water.

3 The ions become evenly distributed. Random movements continue, but there is now no net movement in any particular direction.

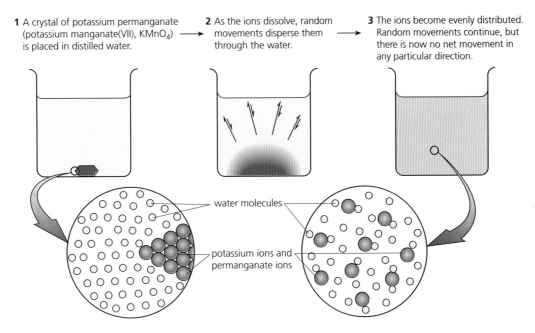

water molecules

potassium ions and permanganate ions

Diffusion in cells

Diffusion across the cell membrane (Figure 2.8) occurs where:

- the plasma membrane is fully permeable to the solute – the lipid bilayer of the plasma membrane is permeable to non-polar substances, including steroids and glycerol, and also oxygen and carbon dioxide in solution, all of which diffuse quickly via this route;
- the pores in the membrane are large enough for a solute to pass through. Water diffuses across the plasma membrane via the protein-lined pores of the membrane (**channel proteins**), and via tiny spaces between the phospholipid molecules. This latter occurs easily where the fluid-mosaic membrane contains phospholipids with unsaturated hydrocarbon tails, for here the hydrocarbon tails are spaced more widely. The membrane is consequently especially 'leaky' to water, for example.

Figure 2.8 Diffusion across the plasma membrane.

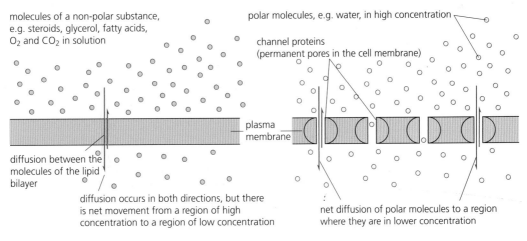

molecules of a non-polar substance, e.g. steroids, glycerol, fatty acids, O_2 and CO_2 in solution

polar molecules, e.g. water, in high concentration

channel proteins (permanent pores in the cell membrane)

plasma membrane

diffusion between the molecules of the lipid bilayer

diffusion occurs in both directions, but there is net movement from a region of high concentration to a region of low concentration

net diffusion of polar molecules to a region where they are in lower concentration

Facilitated diffusion

In facilitated diffusion, a substance that otherwise is unable to diffuse across the plasma membrane does so as a result of its effect on particular molecules present in the membrane. These latter molecules, made of globular protein, form into pores large enough for diffusion – and close up again when that substance is no longer present (Figure 2.9). In facilitated diffusion, the energy comes from the kinetic energy of the molecules involved, as is the case in all forms of diffusion. Energy from metabolism is not required. Important examples of facilitated diffusion are the movement of ADP into mitochondria and the exit of ATP from mitochondria (page 101).

DL
www

Activity 2.3 An analysis of diffusion rates

A* Extension 2.1: Movement of water across the plasma membrane

Figure 2.9 Facilitated diffusion.

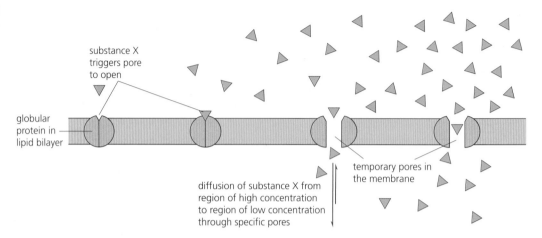

4 Distinguish between *diffusion* and *facilitated diffusion*.

Osmosis – a special case of diffusion

Osmosis is a special case of diffusion. It is the diffusion of water molecules across a membrane which is permeable to water. Since water makes up 70–90% of living cells and cell membranes are such 'partially permeable' membranes, osmosis is very important in biology.

Firstly, why does osmosis happen?

Dissolved substances attract a group of polar water molecules (page 3) around them. The forces holding water molecules in this way are weak chemical bonds, including **hydrogen bonds**. Consequently, the tendency for random movement by these dissolved substances and their

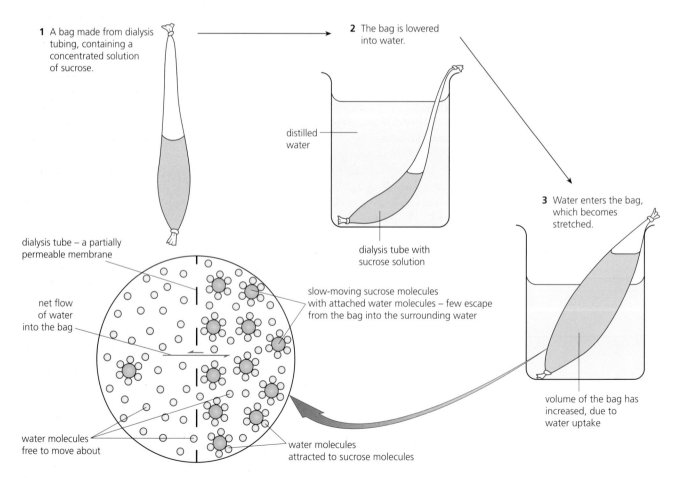

Figure 2.10 Osmosis.

5 What is meant by 'free' water molecules? What is the alternative state of water?

6 Draw a diagram of two adjacent cells, in one of which the cytoplasm contains a more concentrated solution of solute molecules than in the other. Annotate your diagram to show why and in which direction osmosis occurs.

Activity 2.4:
Demonstrating osmotic pressure

A* Extension 2.2:
Osmosis in operation

surrounding water molecules is restricted. Organic substances like sugars, amino acids, polypeptides and proteins, and inorganic ions like Na^+, K^+, Cl^- and NO_3^-, have this effect on the water molecules around them.

The stronger the solution (that is, the more solute dissolved per volume of water), the larger the number of water molecules that are slowed up and held almost stationary. So, in a very concentrated solution, very many more of the water molecules have restricted movement than in a dilute solution. On the other hand, in pure water, all of the water molecules are free to move about randomly, and do so.

When a solution is separated from water (or a more dilute solution) by a membrane permeable to water molecules (such as the plasma membrane), water molecules free to move tend to diffuse, while dissolved molecules and their groups of water molecules move very much less, if at all. So there is a net flow of water into a concentrated solution, from water or a weaker solution, across the membrane. The membrane is described as **partially permeable**.

Osmosis is the net movement of water molecules (solvent), from a solution of high concentration of water molecules to a region of lower concentration of water molecules, across a selectively permeable membrane.

2 Movement by active transport

We have seen that diffusion is due to random movements of molecules, and occurs spontaneously, from a high to a low concentration. However, many of the substances required by cells have to be absorbed from a weak external concentration and taken up into cells that contain a higher concentration. Uptake against a concentration gradient cannot occur by diffusion. Instead, it requires a source of energy to drive it. This type of uptake is known as **active transport**.

In active transport, metabolic energy produced by the cell, held as **adenosine triphosphate (ATP)**, is used to drive the transport of molecules and ions across cell membranes. Active transport has characteristic features distinctly different from those of movement by diffusion.

- **Active transport can occur against a concentration gradient** – that is, from a region of low concentration to a region of higher concentration.
  The cytoplasm of a cell normally holds some reserves of substances valuable in metabolism, like nitrate ions in plant cells, or calcium ions in muscle fibres. The reserves of useful molecules and ions do not escape; the cell membrane retains them inside the cell. Yet when more of these or other useful molecules or ions become available for uptake, they too are actively absorbed into the cells. This happens even though the concentration outside is lower than inside.
- **Active uptake is highly selective.** For example, in a situation where potassium chloride (K^+ and Cl^- ions) is available to an animal cell, K^+ ions are more likely to be absorbed, since they are needed by the cell. Where sodium nitrate (Na^+ and NO_3^- ions) is available to a plant cell, it is likely that more of the NO_3^- ions are absorbed than the Na^+, since this too reflects the needs of the cell.
- **Active transport involves special molecules of the membrane called 'pumps'.** The pump molecule picks up particular ions and molecules and transports them to the other side of the membrane, where they are then released. The pump molecules (also known as **carrier proteins**) are globular proteins that span the lipid bilayer (Figure 2.1). Movements by these pump molecules require reaction with ATP; this reaction supplies metabolic energy to the process. Most membrane pumps are specific to particular ions and molecules and this is the way selective transport is brought about. If the pump molecule for a particular substance is not present, the substance will not be transported.

Active transport is a feature of most living cells. We meet examples of active transport in the active uptake of ions by plant roots, in the mammalian gut where absorption occurs, in the kidney tubules where urine is formed, and in nerve fibres where an impulse is propagated.

The protein pumps of plasma membranes are of different types. Some transport a particular molecule or ion in one direction (Figure 2.11), while others transport two substances (like Na^+ and K^+) in opposite directions (Figure 2.12). Occasionally, two substances are transported in the same direction – for example, Na^+ and glucose during the absorption of glucose in the small intestine.

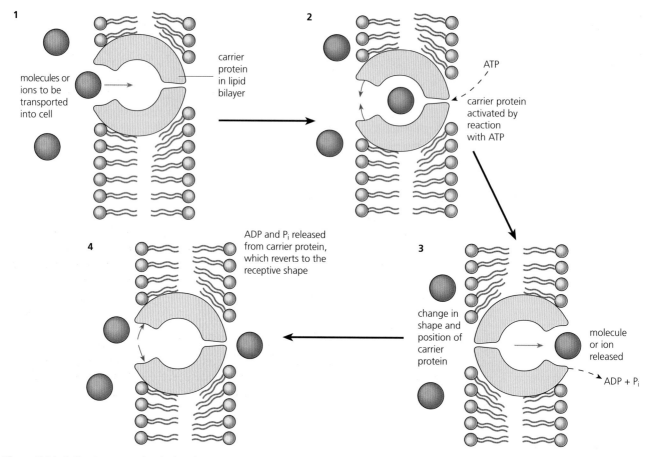

Figure 2.11 Active transport of a single substance.

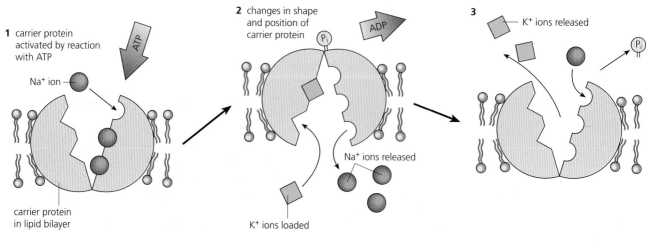

Figure 2.12 The sodium–potassium ion pump.

DL
www
Activity 2.5: Ion absorption in the presence of dinitrophenol (DNP)

7 Samples of five plant tissue discs were incubated in dilute sodium chloride solution at different temperatures. After 24 hours it was found that the uptake of ions from each solution was as shown in the table.

	Uptake of sodium ions/arbitrary units	Uptake of chloride ions/arbitrary units
tissue at 5 °C	80	40
tissue at 25 °C	160	80

Suggest how absorption of sodium chloride occurred, giving your reasons.

3 Movement by bulk transport

Another mechanism of transport across the plasma membrane is known as **bulk transport**. It occurs through the movement of **vesicles** of matter (solids or liquids) across the membrane, by processes known generally as cytosis. Uptake is called **endocytosis** and export is **exocytosis**.

The strength and flexibility of the fluid mosaic membrane makes this activity possible. Energy from metabolism (ATP) is also required to bring it about. For example, when solid matter is being taken in (**phagocytosis**), part of the plasma membrane at the point where the vesicle forms is pulled inwards and the surrounding plasma membrane and cytoplasm bulge out. The matter thus becomes enclosed in a small vesicle.

8 By what processes do:
 a oxygen
 b water
 c glucose
 d bacteria
 enter a living white blood cell?

In the human body, there are a huge number of phagocytic cells, which are called the **macrophages**. The macrophages engulf the debris of damaged or dying cells and dispose of it (phagocytosis means 'cell eating'). For example, we break down about 2×10^{11} red cells each day! This number are ingested and disposed of by macrophages, every 24 hours.

Bulk transport of fluids is referred to as **pinocytosis** (Figure 2.13).

Figure 2.13 Transport by cytosis.

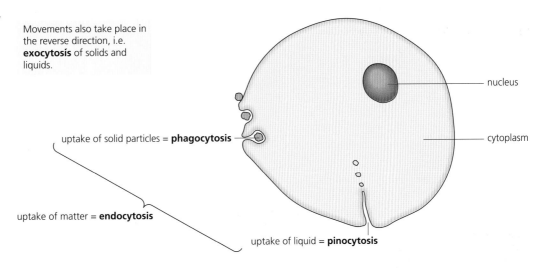

Movements also take place in the reverse direction, i.e. **exocytosis** of solids and liquids.

nucleus

cytoplasm

uptake of solid particles = **phagocytosis**

uptake of matter = **endocytosis**

uptake of liquid = **pinocytosis**

■ Investigation: effect of heat on the permeability of plasma membranes

The cells of beetroot contain an intensely red, water-soluble pigment in their interior – it is found in the large central vacuole of each cell. The pigment may escape in sufficient quantities to be detected in the aqueous medium around the tissue. This occurs if harmful external conditions are applied to beetroot tissue. For example, let us consider the effect of externally applied heat energy. One effect of heat is to denature proteins, and this applies to the proteins of membranes as well as those elsewhere in cells. Once membrane proteins have been denatured, the integrity of the lipid bilayer of the plasma membranes may also be compromised. Escape of the vacuole contents will indicate this has happened. So, in effect we can experimentally investigate the approximate temperature at which membrane proteins are seriously denatured.

It is also possible to adapt this technique to investigate the effect of chemical substances (for example, strong solutions of ions, or organic solvents such as alcohol) on the permeability of membranes.

Aim: To determine the temperature at which the beetroot plasma membrane is denatured by heat.
Risk assessment: Good laboratory practice is sufficient to avoid a hazard.
Method: The first step is to prepare washed beetroot tissue cylinders, about 3 cm long and 0.5 cm in diameter (Figure 2.14). Ten cylinders should be cut.

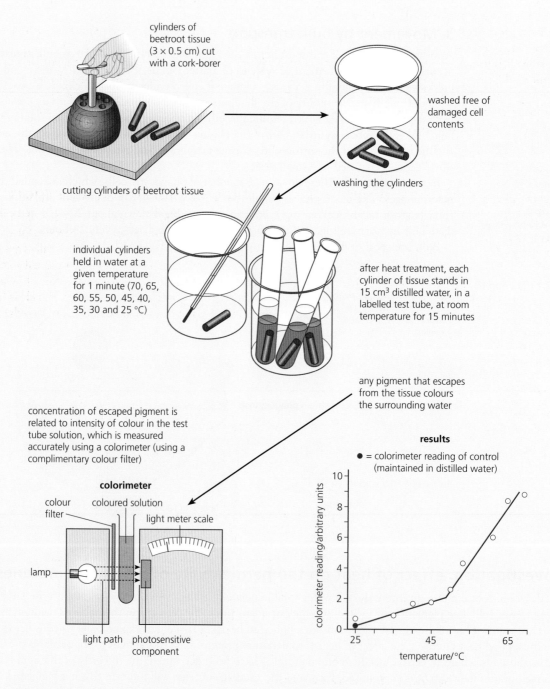

Figure 2.14 Investigating membrane permeability.

Submerge one cylinder in a water bath at 70 °C for 1 minute. Then withdraw it and place it in 15 cm³ of distilled water in a test tube (labelled 'treatment at 70 °C') at room temperature for 15 minutes. After this, the tissue cylinder should be removed and discarded.

Cool the water bath to 65 °C and repeat the process with a second cylinder – heat-treat it, allow it to stand in a tube of fresh distilled water (labelled 'treatment at 65 °C') for 15 minutes, and then discard it.

Repeat, using heat treatments with water that is 5 °C cooler each time. Make 25 °C the lowest temperature treatment.

The distilled water in the test tubes becomes coloured by any pigment that has escaped from the tissue cylinders as a result of heat treatment. Measurement of the pigment loss from the tissue into the test tube solutions is undertaken using a colorimeter containing a complementary colour filter (so, for a red solution, a blue filter is required). Set the scale zero using the solvent (distilled water) only.

Data presentation: A graph of the colorimeter readings plotted against temperature will provide a curve from which the approximate temperature at which the escape of pigment was accelerated can be read (Figure 2.14).

Conclusion: It is possible to deduce the approximate temperature at which membrane proteins are seriously denatured. If the value is higher than you might expect (compared with the temperatures that denature typical plant enzymes, for example), comment on possible explanations by analysis of your experimental method. How quickly do the cells at the centre of your tissue cylinders experience the water temperature to which they are treated, for example?

Evaluation: What improvements to your technique might enhance the accuracy of your method?

HSW 2.2: Criteria 2, 3, 4, 5 and 8 – A physiological investigation of plasma membrane permeability in beetroot

9 Using your knowledge of the structure and composition of the plasma membrane, predict the effects that exposure of the membrane to an organic solvent such as ethanol will have.

Gas exchange – a case study in membrane transport

Respiration, the cell process by which energy is transferred for all the activities of life, is continuous in all living things. Typically, animal and plant cells respire **aerobically** – they take in oxygen from their environment and give out carbon dioxide by a process called gas exchange. Gas exchange in cells occurs by diffusion. For example, in cells respiring aerobically there is a higher concentration of oxygen outside the cells than inside, and so there is a net inward diffusion of oxygen (Figure 2.15). Conversely, the concentration of carbon dioxide is higher inside. Wherever a difference in concentrations occurs, there is net diffusion from the higher to a lower concentration.

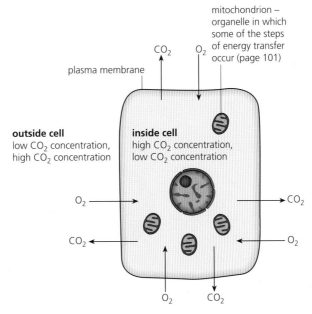

Figure 2.15 Gas exchange in a respiring animal cell.

What speeds up diffusion?

The rate of diffusion varies, depending on a number of factors. For example, an increase in temperature speeds up the random movements of molecules and so an increase in the temperature speeds up diffusion. However, in living things, temperature is normally relatively low and fairly constant. In practice, temperature is not an important factor in deciding the diffusion rate here. Within biological systems, three factors effectively determine the rate of diffusion in practice.

1   **The surface area available for gas exchange** in the organism, called the respiratory surface – the greater this surface area, the greater the rate of diffusion. In a single cell, the respiratory surface is the whole plasma membrane.
2   **The difference in concentration** – a rapidly respiring organism will have a very much lower concentration of oxygen in the cells and a higher than normal concentration of carbon dioxide. The greater the gradient in concentration across the respiratory surface, the greater the rate of diffusion.
3   **The length of the diffusion path** – the shorter the diffusion path, the greater the rate of diffusion, so the respiratory surface must be as thin as possible.

■ Extension: The rate of diffusion

The relationship of the three factors affecting the rate of diffusion, listed above, is summarised by Fick's Law. This states that:

$$\text{rate of diffusion} \propto \frac{\text{surface area} \times \text{difference in concentration}}{\text{length of diffusion pathway}}$$

The respiratory surface – size and shape of organisms

The size and shape of an organism influences its gas exchange. The amount of gas an organism needs to exchange is largely proportional to its volume (the bulk of respiring cells), but the amount of exchange that can occur is proportional to its surface area over which diffusion takes place. For example, in an organism that consists of a single cell, the surface area is large in relation to the amount of cytoplasm it contains. Here the surface of the cell is sufficient for efficient gas exchange because the sites where respiration occurs in the cytoplasm are never very far from the surface of the cell. The surface-area-to-volume ratio is very high for single-celled organisms, and this makes for efficient gas exchange.

The geometric 'organisms' in Figure 2.16 illustrate how the surface-area-to-volume ratio changes as the size of an organism increases. Increasing size lowers the surface area per unit of

Figure 2.16 Size and shape, and surface-area-to-volume ratios.

Shape of organism	'spheroid'		'cuboid'		'thin and flat'	
Size	small	large	small	large	small	large
Dimensions/mm (diameter/mm)	(1)	(4)	1 × 1 × 1	4 × 4 × 4	2 × 1 × 0.5	16 × 8 × 0.5
Volume/mm³	0.5	33.5	1	64	1	64
Surface area/mm²	3	50	6	96	7	280
SA/V ratio	3/0.5 = 6	50/33.5 = 1.5	6/1 = 6	96/64 = 1.5	7/1 = 7	280/64 = 4.4

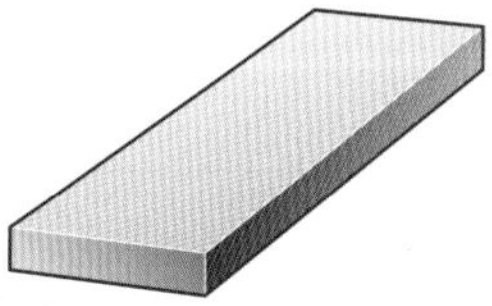

volume of the whole structure – that is, the larger the object, the smaller its surface-area-to-volume ratio. However, the shape of an organism is also important to diffusion of gases in and out. A thin and flat shape – such as that of the leaves of a plant, the fronds of seaweed and the body of a flatworm – has a large surface-area-to-volume ratio and therefore gas exchange is extremely efficient.

Amoeba, a large, single-celled animal (protozoan) living in pond water and feeding on the tiny protozoa around it. Food is taken into food vacuoles. Gases are exchanged over the whole body surface.

Size = about 400 μm

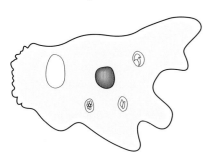

Ulva, the sea lettuce, an anchored or free-floating seaweed. It floats near the surface of water and photosynthesises in the light. Gases are exchanged over the whole body surface.

Size = about 5–15 cm long, about 30–35 μm thick

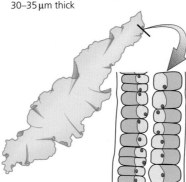

Dugesia tigrina, a free-living flatworm found in ponds under stones or leaves or gliding over the mud. It feeds on smaller animals and fish eggs. It is a very thin animal that exchanges gases over the whole body surface.

Size = about 20 mm

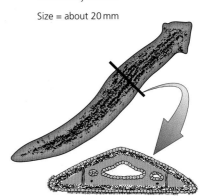

Figure 2.17 Organisms in which gas exchange occurs through their external surface.

specialised respiratory systems have:
- a very large area in a small space
- a ventilation mechanism to maintain the supply of air or oxygenated water

and may have:
- an internal transport system to deliver oxygen to tissues
- a respiratory pigment to improve the efficiency of transport

gills:
- internal or external
- compact, but with a large surface area
- blood circulates between gills and body
- water forced over gills by muscle action

lungs:
- internal, with a large surface area
- blood circulates between lungs and body
- air drawn into and out of lungs by muscle action

tracheae:
- internal tubes that divide and reach the cells of the body
- air moves mostly by diffusion but may be pumped by simple bellows mechanism

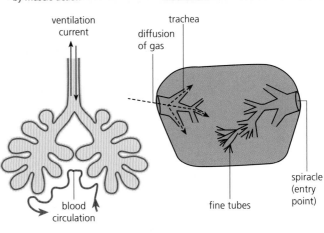

External gills are present in young tadpoles; internal gills are present in older tadpoles and in fish.

Lungs are typical of mammals and other vertebrates.

Tracheae are present in many insects.

Figure 2.18 Examples of specialised respiratory systems in animals.

DL
www
Activity 2.6:
Investigating size and
surface-area-to-volume
ratios

Respiratory systems in large animals

In larger animals, many of their cells are too far from the body surface to receive enough oxygen by diffusion alone. In addition, many of these animals have developed an external surface that provides protection to the body, such as an impervious, water-tight outer covering, or tough or hardened skin. These outer surfaces are no longer suitable for gas exchange, and the organism requires an alternative respiratory surface. Active organisms have an increased metabolic rate, too, and the demand for oxygen in their cells is higher than in sluggish and inactive organisms. So we find that for many reasons, larger, active animals have specialised organs for gas exchange.

Specialised, efficient respiratory surfaces in animals take various forms (Figure 2.18), such as gills in fish, or lungs in mammals, or a tubular system in many insects that carries air to the most actively respiring organs. All these systems provide a large, thin surface area, suitable for gas exchange. In addition, conditions for diffusion are often improved by three refinements:

1 **a ventilation mechanism** – a pumping mechanism that moves the respiratory medium (water or air) over the gills or into and out of the lungs or tubes, thereby maintaining the concentration gradient for diffusion
2 **a blood circulatory system** – a means of accelerating the removal of dissolved oxygen from the respiratory surface as soon as it has diffused in, thereby maintaining the concentration gradient
3 **a respiratory pigment** which increases the gas-carrying ability of the blood – for example, our blood contains red cells packed with the respiratory pigment haemoglobin.

We can now consider the human lungs as structures adapted for rapid gas exchange.

The working lungs of mammals

The structure of the human thorax is shown in Figure 2.19. Lungs are housed in the **thorax**, an air-tight chamber formed by the rib-cage and its muscles (**intercostal muscles**), with a domed floor, which is the **diaphragm**. The diaphragm is a sheet of muscle attached to the body wall at the base of the rib-cage, separating thorax from abdomen. The internal surfaces of the thorax are lined by the **pleural membrane**, which secretes and maintains pleural fluid. Pleural fluid is a lubricating liquid derived from blood plasma that protects the lungs from friction during breathing movements.

The lungs connect with the pharynx at the rear of the mouth by the **trachea**. Air reaches the trachea from the mouth and nostrils, passing through the larynx ('voice box'). Entry into the larynx is via a slit-like opening, the glottis. Above is a cartilaginous flap, the **epiglottis**. Glottis and epiglottis work to prevent the entry of food into the trachea. The trachea initially runs beside the oesophagus (food pipe). Incomplete rings of cartilage in the trachea wall prevent collapse under pressure from a large bolus (ball of food) passing down the oesophagus.

The trachea then divides into two **bronchi**, one to each lung. Within the lungs the bronchi divide into smaller bronchioles. The finest bronchioles end in air sacs (**alveoli**). The walls of bronchi and larger bronchioles contain smooth muscle, and are also supported by rings or tiny plates of cartilage, preventing collapse that might be triggered by a sudden reduction in pressure that occurs with powerful inspirations of air.

Ventilation of the lungs

Air is drawn into the alveoli when the air pressure in the lungs is lower than atmospheric pressure, and it is forced out when pressure is higher than atmospheric pressure. Since the thorax is an air-tight chamber, pressure changes in the lungs occur when the volume of the thorax changes. The ventilation mechanism of the lungs is summarised in Table 2.3.

10 List three characteristics of an efficient respiratory surface and give a reason why each influences diffusion.

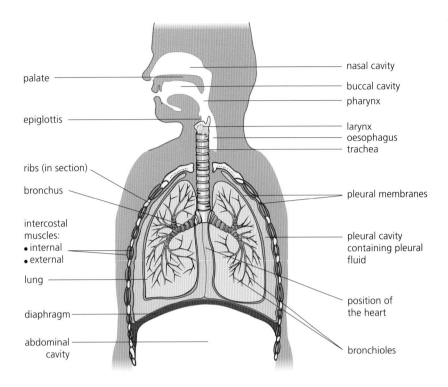

Figure 2.19 The structure of the human thorax.

Table 2.3 The mechanism of lung ventilation – a summary

Inspiration (inhalation)		Expiration (exhalation)
muscles contract, flattening the diaphragm, pushing down on contents of abdomen	**diaphragm**	muscles relax – pressure from abdominal contents pushes diaphragm back up into dome shape
contract, moving rib-cage up and out	**external intercostal muscles**	relax
relax	**internal intercostal muscles**	contract, moving rib-cage down and in
volume increases	**volume of thorax cavity**	volume decreases
falls below atmospheric pressure	**air pressure of thorax**	rises above atmospheric pressure
in	**air flow**	out

Alveolar structure and gas exchange

The lung tissue consists of the alveoli, arranged in clusters, each served by a tiny bronchiole (Figure 2.20). Alveoli have elastic connective tissue as an integral part of their walls. A capillary system wraps around the clusters of alveoli. Each capillary is connected to a branch of the pulmonary artery and drained by a branch of the pulmonary vein. The pulmonary circulation is supplied with deoxygenated blood from the right side of the heart, and oxygenated blood is returned to the left side of the heart to be pumped to the rest of the body.

There are some 700 million alveoli present in our lungs, providing a surface area of about 70 m² in total. This is an area 30–40 times greater than that of the body's external skin.

The wall of an alveolus is one cell thick, formed by pavement **epithelium**. An epithelium is a sheet of cells bound strongly together, covering internal or external surfaces of multicellular organisms. Lying very close is a capillary, its wall also composed of a single, flattened (endothelium) cell. The combined thickness of walls separating air and blood is typically 5 μm thick. The capillaries are extremely narrow, just wide enough for red cells to squeeze through, so red cells are close to or in contact with the capillary walls.

Activity 2.7: Wider reading – 'Gas exchange in the lungs'

Blood arriving in the lungs is low in oxygen but high in carbon dioxide. As blood flows past the alveoli, gas exchange occurs by diffusion. Oxygen dissolves in the surface film of water, then diffuses across into the blood plasma and into the red cells where it combines with haemoglobin to form oxyhaemoglobin. At the same time, carbon dioxide diffuses from the blood into the air in the alveolus. Table 2.4 shows how the composition of air varies during breathing. Figure 2.20 illustrates how the fine structure of the lung tissue facilitates gas exchange.

A* Extension 2.3:
Protection of the lungs

Table 2.4 The composition of air in the lungs.

	Percentage of each gas present/%		
	inspired air	alveolar air	expired air
Oxygen	20	14	16
Carbon dioxide	0.04	5.5	4.0
Nitrogen	79	81	79
Water vapour	variable	saturated	saturated

11 Suggest why build up of carbon dioxide concentration in the blood of a mammal would be harmful.

Figure 2.20 Gas exchange in the alveoli.

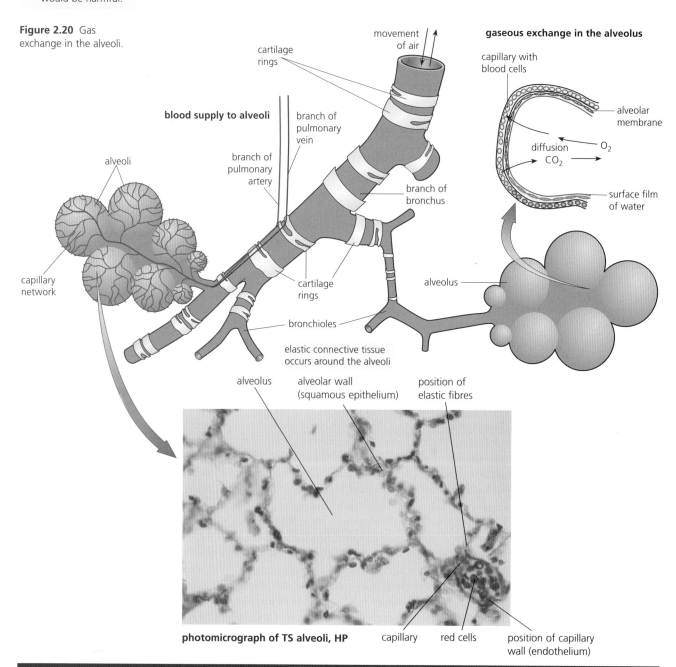

photomicrograph of TS alveoli, HP

Activity 2.8: The alveoli and efficient gas exchange

■ **Extension: Efficiency of lungs as organs of gas exchange**

Air flow in the lungs of mammals is tidal, in that air enters and leaves by the same route. Consequently there is a residual volume of air that cannot be expelled. Incoming air mixes with and dilutes the residual air. The effect of this is that air in the alveoli contains significantly less oxygen than the atmosphere outside (Table 2.4). Nevertheless, the lungs are efficient organs of gas exchange.

2.2 Proteins and enzymes

Proteins make up about two-thirds of the total dry mass of a cell. They differ from carbohydrates and lipids in that they contain the element nitrogen, and usually the element sulphur, as well as carbon, hydrogen and oxygen. **Amino acids** are the molecules from which peptides and proteins are built – typically several hundred or even thousands of amino acid molecules are combined together to form a protein. Incidentally, the terms 'polypeptide' and 'protein' can be used interchangeably, but when a polypeptide is about 50 or more amino acid residues long it is generally agreed to be a protein.

Once the chain is constructed, a protein takes up a specific shape. Shape matters with proteins – their shape is closely related to their function. This is especially the case in proteins that are **enzymes**, as we shall shortly see.

Amino acids – the building blocks of peptides

As their name implies, amino acids carry two groups:

■ an amino group ($-NH_2$)
■ an **organic acid group** (carboxyl group, $-COOH$).

These groups are attached to the same carbon atom in the amino acids, which get built up into proteins. Also attached here is a side chain part of the molecule, called an **R group** (Figure 2.21).

Figure 2.21 The structure of amino acids.

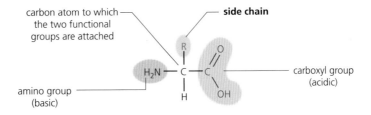

The 20 different amino acids that make up proteins in cells and organisms differ in their side chains (R groups). Below are three examples.

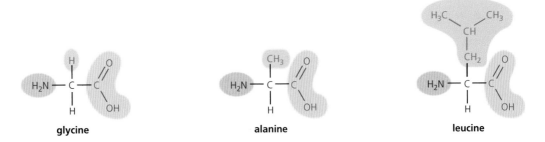

glycine   alanine   leucine

Some amino acids have an additional $-COOH$ group in their side chain (= acidic amino acids).
Some amino acids have an additional $-NH_2$ group in their side chain (= basic amino acids).

While there very many different types of amino acids, the proteins found in living things are built from just 20 different amino acids, present in differing proportions. All we need to note about these molecules is that their R groups are all very different. Consequently, all 20 amino acids (and proteins containing them) have different chemical characteristics.

Peptide linkages

In the presence of the required enzyme, two amino acids combine together with the loss of water to form a dipeptide. This is one more example of a **condensation reaction**. The amino group of one amino acid reacts with the carboxyl group of the other, forming a **peptide linkage**.

A further condensation reaction between the dipeptide and another amino acid results in a tripeptide. In this way, long strings of amino acid residues, joined by peptide linkages, are formed (Figure 2.22). Thus, peptides or protein chains are assembled, one amino acid at a time, in the presence of specific enzymes (page 77).

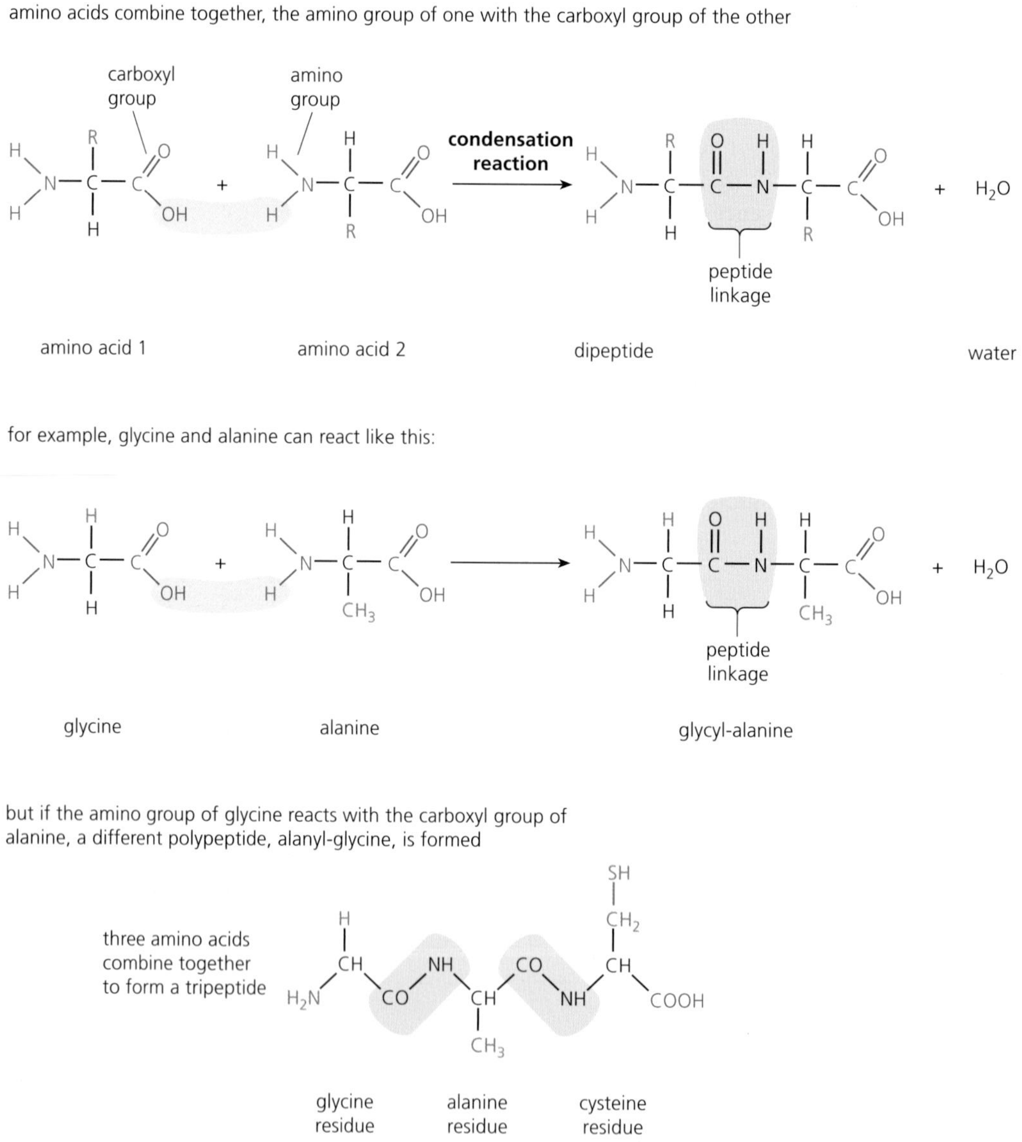

Figure 2.22 Peptide linkage formation.

amino acids combine together, the amino group of one with the carboxyl group of the other

for example, glycine and alanine can react like this:

but if the amino group of glycine reacts with the carboxyl group of alanine, a different polypeptide, alanyl-glycine, is formed

three amino acids combine together to form a tripeptide

The structure of proteins

We have already noted that the shape of a protein molecule is critical in determining the properties and role that protein has in a cell. In fact, there are four levels of structure to a protein: primary, secondary, tertiary and quaternary.

A* Extension 2.4:
Potential variations in
the structure of a
polypeptide

1 **The primary structure of a protein** is the sequence of the amino acids in its molecules. Proteins differ in the variety, number and order of their constituent amino acids. The order of amino acids in the polypeptide chain is controlled by the coded instructions stored in the DNA of the chromosomes in the nucleus (page 75). Just changing one amino acid in the sequence of a protein may alter its properties completely. This sort of 'mistake' or **mutation** does happen (page 79).

2 **The secondary structure of a protein** develops when parts of the polypeptide chain take up a particular shape, immediately after formation at the ribosome. Parts of the chain become folded or twisted, or both, in various ways. The most common shapes are formed either by coiling to produce an α helix or folding into β sheets. These shapes are permanent, held in place by hydrogen bonds (page 2).

α **helix** (rod-like)

β **sheets**

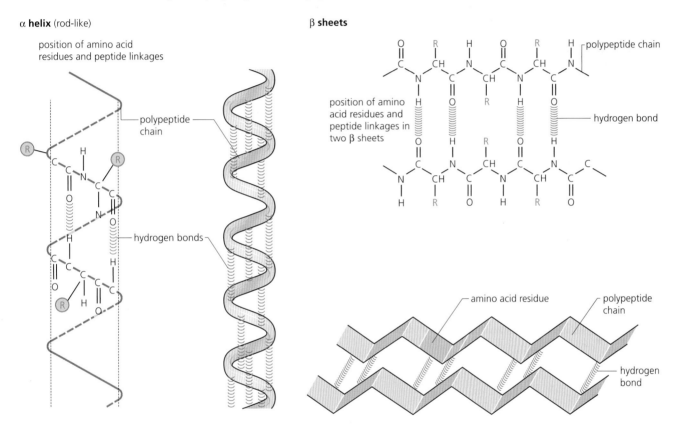

Figure 2.23 The secondary structure of protein.

3 **The tertiary structure of a protein** is the precise, compact structure, unique to that protein, that arises when the molecule is further folded and held in a particular complex shape. This shape is made permanent by four different types of bonding, established between adjacent parts of the chain (Figure 2.24). The primary, secondary and tertiary structure of the protein lysozyme is shown in Figure 2.25.

Some proteins take up a tertiary structure which is a long, much-coiled chain, and are called **fibrous proteins**. Examples of fibrous proteins are fibrin, a blood protein involved in the clotting mechanism, collagen, a component of bone and tendons, and keratin, found in hair, horn and nails.

Other proteins take up a tertiary structure that is more spherical, and are called **globular proteins**. Enzymes are typically globular proteins.

4 **The quarternary structure of protein** arises when two or more proteins become held together, forming a complex, biologically active molecule. An example is the respiratory pigment haemoglobin, found in red cells. This molecule consists of four polypeptide chains held around a non-protein haem group, in which an atom of iron occurs.

Figure 2.24 Cross-linking within a polypeptide.

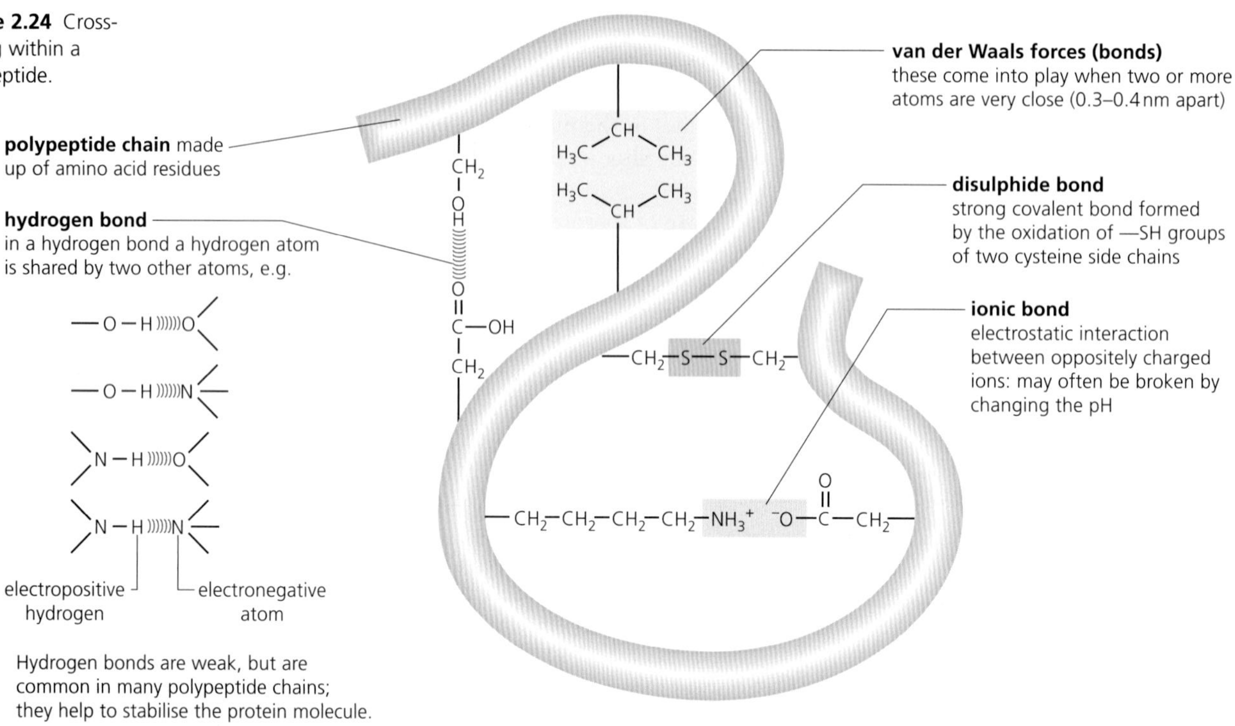

polypeptide chain made up of amino acid residues

hydrogen bond
in a hydrogen bond a hydrogen atom is shared by two other atoms, e.g.

—O—H))))))O
—O—H))))))N
N—H))))))O
N—H))))))N

electropositive hydrogen — electronegative atom

Hydrogen bonds are weak, but are common in many polypeptide chains; they help to stabilise the protein molecule.

van der Waals forces (bonds)
these come into play when two or more atoms are very close (0.3–0.4 nm apart)

disulphide bond
strong covalent bond formed by the oxidation of —SH groups of two cysteine side chains

ionic bond
electrostatic interaction between oppositely charged ions: may often be broken by changing the pH

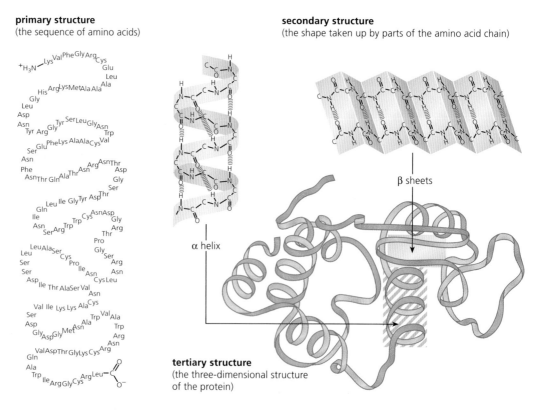

Figure 2.25 Lysozyme, primary, secondary and tertiary structure.

primary structure
(the sequence of amino acids)

secondary structure
(the shape taken up by parts of the amino acid chain)

β sheets

α helix

tertiary structure
(the three-dimensional structure of the protein)

What determines the 3-D shape of a protein?

When a protein loses its three-dimensional shape, we say it has been **denatured**. When this happens to small proteins, it is found that they sometimes revert back to their former shape once the conditions that triggered denaturation are removed. This observation led to the idea that it is simply the amino acid sequence of a protein that decides its tertiary structure. This may well be true for many polypeptides and small proteins. However, in most proteins within the cell environment, folding is a speedy process in which some accessory proteins, including enzymes, are normally involved. These may determine the shape as much as, or more than, the amino acid sequence does.

Roles of proteins

Many proteins have a structural role – in membranes around the cell, for example (and in many internal structures, known as cell organelles). Another protein with a structural function is the fibrous protein collagen, found in hair, tendon, bones and teeth. Certain membrane proteins are 'pumps' that transport molecules across the membrane. Other proteins are carrier molecules, involved in bulk transport of essential resources (page 51). There are important proteins in the blood, too, and some are antibodies involved in combating disease and invasion by bacteria. Some hormones are also proteins.

However, very many cell proteins are **enzymes** – some are free-floating in the liquid part of the cytoplasm, others are built into membranes. Here, we shall concern ourselves only with this particular role of proteins.

Enzymes – biological catalysts

A* Extension 2.5: Can a reaction occur without an enzyme?

Most chemical reactions do not occur spontaneously. In a laboratory or in an industrial process, chemical reactions may be made to occur by applying high temperatures, high pressures, extremes of pH, and by maintaining high concentrations of the reacting molecules. If these drastic conditions were not applied, very little of the chemical product would be formed. On the other hand, in cells and organisms, many chemical reactions occur simultaneously, at extremely low concentrations, at normal temperatures, and under the very mild, almost neutral, aqueous conditions we find in cells.

How are these reactions brought about?
It is the presence of enzymes in cells and organisms that enables these reactions to occur at incredible speeds, in an orderly manner, yielding products that the organism requires, when they are needed. Sometimes reactions happen even though the reacting molecules are present in very low concentrations. Enzymes are truly remarkable molecules.

So, enzymes are biological catalysts made of protein. By 'catalyst', we mean a substance that speeds up the rate of a chemical reaction. The general properties of **catalysts** are that they:

- are effective in small amounts
- remain unchanged at the end of the reaction.

Another important property of enzymes and catalysts is that they speed up the rate at which an equilibrium position is reached. In a reversible reaction:

$$A + B \rightleftharpoons C + D$$

A and B react to form C and D. This is a reversible reaction as shown by the sign $\rightleftharpoons$. As soon as C and D start to accumulate, some will react to form A and B. The reversible reaction reaches an equilibrium point when the rate of the forward reaction equals the rate of the reverse reaction. Most enzyme-catalysed reactions are reversible; the presence of the enzyme for these reactions means the equilibrium position is reached quickly.

Where do enzymes operate?
Some enzymes are exported from cells, such as the digestive enzymes. Enzymes like these, that are parcelled up and secreted and work externally, are called extracellular enzymes. However, very many enzymes remain within the cells and work there. These are the intracellular enzymes. They are found inside the sub-structures within cells (known as organelles), in the membranes of organelles, in the fluid medium around the organelles, and in the plasma membrane.

What roles do enzymes have in organisms?

There is a huge array of chemical reactions that go on in cells and organisms – collectively, the chemical reactions of life are called **metabolism**. Since each reaction of metabolism can only occur in the presence of a specific enzyme, we know that if an enzyme is not present then the reaction it catalyses cannot occur.

Many enzymes are always present in cells and organisms, but some enzymes are produced only under particular conditions or at certain stages. By making some enzymes and not others, cells can control what chemical reactions happen in the cytoplasm. Later in this chapter we shall see how protein synthesis (and therefore enzyme production) is controlled by the cell nucleus.

The enzyme–substrate complex

Enzymes (E) work by binding to a specific substance, known as their substrate molecule (S), at a specially formed pocket in the enzyme, called the **active site** (Figure 2.26). As the enzyme and substrate form a complex (E–S), this immediately breaks down to form the products (Pr), plus the unchanged enzyme.

$$E + S \rightleftharpoons E\text{–}S \rightleftharpoons Pr + E$$

Figure 2.26 The enzyme–substrate complex and the active site.

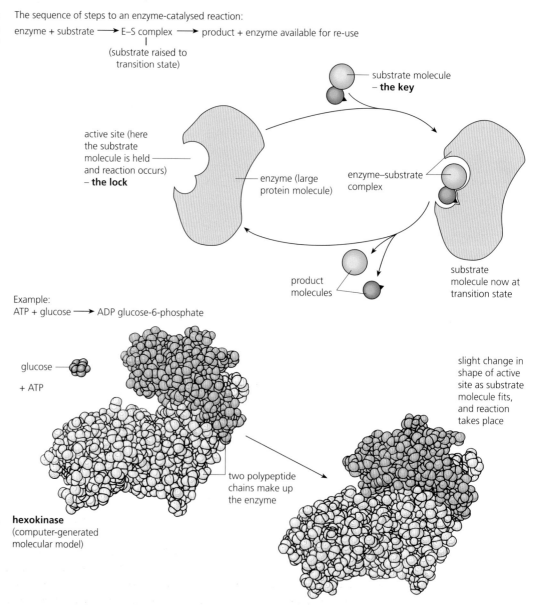

The sequence of steps to an enzyme-catalysed reaction:

enzyme + substrate ⟶ E–S complex ⟶ product + enzyme available for re-use

(substrate raised to transition state)

substrate molecule – **the key**

active site (here the substrate molecule is held and reaction occurs) – **the lock**

enzyme (large protein molecule)

enzyme–substrate complex

product molecules

substrate molecule now at transition state

Example:
ATP + glucose ⟶ ADP glucose-6-phosphate

glucose

+ ATP

slight change in shape of active site as substrate molecule fits, and reaction takes place

two polypeptide chains make up the enzyme

hexokinase (computer-generated molecular model)

To emphasise this, the binding of enzyme and substrate is referred to the 'lock and key' hypothesis of enzyme action. The enzyme is the lock, and the substrate is a key that fits the lock (Figure 2.26). Enzymes are typically large molecules and most substrate molecules are quite small by comparison. Even when the substrate molecules are very large, such as certain macromolecules like the polysaccharides, only one bond in the substrate is in contact with the enzyme active site. The active site takes up a relatively small part of the total volume of the enzyme.

Enzymes lower the energy of activation

As molecules react they become unstable, high-energy intermediates, but only momentarily, while in transition state. Effectively, the products are formed immediately. The minimum amount of energy needed to raise substrate molecules to their transition state is a called the **activation energy**. This is the energy barrier that has to be overcome before the reaction can happen. Enzymes work by lowering the amount of energy required to activate the reacting molecules. Of course, the products have a lower energy level than the substrate molecules.

One way to visualise what is going on is the 'boulder on hillside' model (Figure 2.27). Think of a boulder perched on a slope, prevented from rolling down by a small hump in front of it. The boulder represents the substrate, while the hump is the activation energy. The boulder can be pushed over the hump, or the hump can be dug away (that is, the activation energy can be lowered), allowing the boulder to roll down and shatter at a lower level (the shattered rock represents the products).

'boulder on hillside' model of activation energy

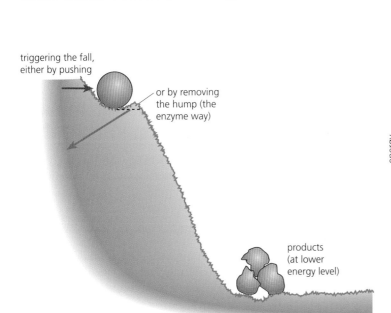

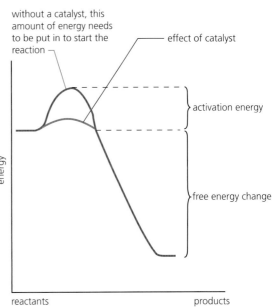

Figure 2.27 Activation energy.

The active site and enzyme specificity

Enzymes are highly specific in their action – they catalyse just one type of reaction or only a small group of highly similar reactions. An enzyme 'recognises' only a small group of substrate molecules, or even just a single type of molecule. This is because the active site where the substrate molecule binds has a precise shape and distinctive chemical properties (meaning the presence of particular chemical groups and bonds). Only particular substrate molecules are attracted to a particular active site and can fit there. All other substrate molecules are unable to fit and so cannot bind.

A* Extension 2.6: What happens at the active site?

12 Explain why the shape of globular proteins that are enzymes is important in enzyme action.

DL
wWw
Activity 2.9: Wider reading – 'Enzymes'

HSW 2.3: Criterion 1 – Models as part of a theory

DL
wWw
Activity 2.10: Demonstrating the effect of catalase on hydrogen peroxide solution

The rate of enzyme-catalysed reactions

Reactions catalysed by enzymes are typically extremely fast, but the rate of an enzyme-catalysed reaction is sensitive to the conditions in which it occurs – we call these the 'environmental conditions'. Any factor that influences an enzyme will alter the rate of the reaction being catalysed, and this can be measured. The way enzyme action is influenced by external conditions may particularly reflect the fact that enzymes are proteins (the effects of temperature and pH are examples). Other factors illustrate in particular the theory that an enzyme works by forming a short-lived enzyme–substrate complex. The effects of substrate and enzyme concentrations are examples, and we will examine these factors next.

1 The effect of different concentrations of substrate

The effect of substrate concentration can be investigated using an enzyme called **catalase**. This enzyme catalyses the breakdown of hydrogen peroxide:

$$2H_2O_2 \xrightarrow{\text{catalase}} 2H_2O + O_2$$

Catalase occurs very widely in living things; it functions as a protective mechanism for the delicate biochemical machinery of cells. This is because hydrogen peroxide is a common by-product of reactions of metabolism, but it is also a very toxic substance (a very powerful oxidising agent). Catalase inactivates hydrogen peroxide as it forms, before damage can occur.

In measuring the rate of enzyme-catalysed reactions, 'rate' may be taken as the amount of substrate that has disappeared from a reaction mixture, or the amount of product that has accumulated, in a unit of time. Working with catalase, it is convenient to measure the rate at which the product (oxygen) accumulates – the volume of oxygen that has accumulated at 30-second intervals is recorded (Figure 2.28).

Over a period of time, the initial rate of reaction is not maintained but, rather, falls off quite sharply. This is typical of enzyme actions studied outside their location in the cell. The fall-off can be for a number of reasons, but most commonly it is because the concentration of the substrate in the reaction mixture has fallen. Consequently, it is the initial rate of reaction that is measured. This is the slope of the tangent to the curve in the initial stage of reaction.

We can see from the graph in Figure 2.29 that when the initial rates of reaction are plotted against the substrate concentration, the curve shows two phases. At lower concentrations, the rate increase is in direct proportion to the substrate concentration, but at higher substrate concentrations the initial rate of reaction is constant, and shows no increase with increased concentration. Interpreting this graph, we can say that the enzyme catalase works by forming a short-lived enzyme–substrate complex. At low concentration of substrate, all molecules can find an active site without delay. Effectively, there is an excess enzyme present. Here the rate of reaction is set by how much substrate is present – as more substrate is made available, the rate of reaction increases.

However, at higher substrate concentrations there comes a point where there is more substrate present than enzyme to catalyse it. Now, in effect, substrate molecules have to 'queue up' for access to an active site. Adding more substrate increases the number of molecules awaiting contact with an enzyme molecule; there is now no increase in the rate of reaction.

2 The effect of enzyme concentration

If there is an excess of substrate molecules in a reaction mixture, then the more enzyme that is added the faster will be the rate of reaction. This is the situation in a cell where an enzyme reaction occurs with only a small amount of enzyme present. Any increase in production of that enzyme will lead to an increased rate of reaction, simply because more active sites are made available. As a consequence, in a laboratory investigation of the effect of enzyme concentration we might guess that the relationship between reaction rate (dependent variable – y-axis) and concentration of enzyme (independent variable – x-axis) will be a linear one.

Such an investigation is one of the *core practicals* of your course. The enzyme catalase and the apparatus illustrated in Figure 2.28 can be adapted to investigate the effect of enzyme concentration, since once again the attention is on the initial rate of reaction.

What are the changes in experimental method needed for this study?

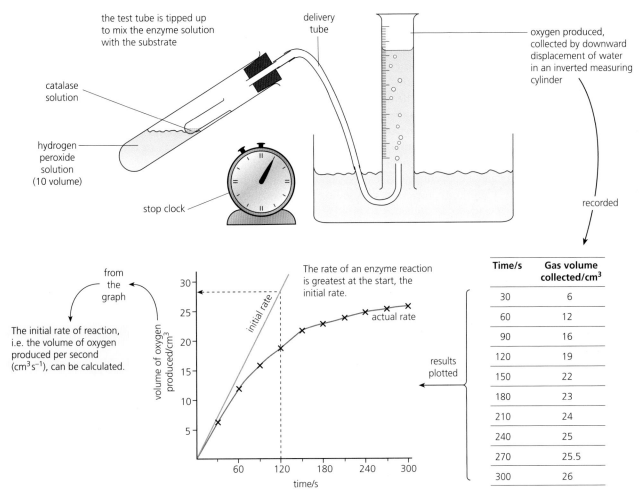

Time/s	Gas volume collected/cm³
30	6
60	12
90	16
120	19
150	22
180	23
210	24
240	25
270	25.5
300	26

Figure 2.28 Measuring the rate of reaction, using catalase.

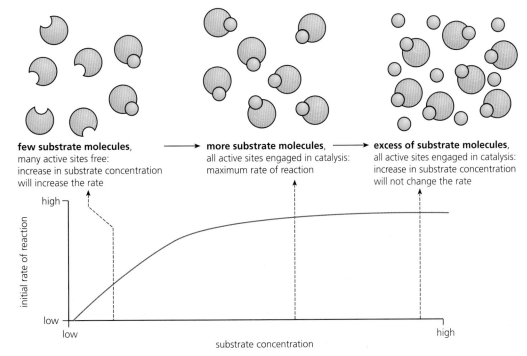

Figure 2.29 The effect of substrate concentration.

Firstly, the same concentration of hydrogen peroxide (the substrate) will need to be used in each investigation here. Hydrogen peroxide solution is normally available as a bench reagent, as a 20-volume solution. It is probably a good idea to select a fairly high concentration (minimal dilution with distilled water).

Can you suggest why?

It would need little dilution with distilled water if a significant and measurable yield of oxygen is to be obtained at the lower concentrations of enzyme you select. After all, it's the initial rate of reaction that you need to measure at each enzyme concentration.

Next, select the range of concentration of enzyme to be used. Yeast cells are a rich source of catalase, so a yeast suspension is a practical alternative to a specially prepared enzyme solution. A range of concentrations may be obtained by making serial dilutions of a stock suspension.

Planning, conducting, concluding and reporting on this investigation touches on a number of *How science works* issues, too (HSW **2.4**).

HSW 2.4: Criteria 2, 3, 4, 5 and 8 – An investigation of the effects of substrate concentration on enzyme action

Activity 2.11: Wider reading – 'Enzymes: fast and flexible'

13 When there is an excess of substrate present in an enzyme-catalysed reaction, explain the effects on the rate of reaction of increasing the concentration of:
   a the substrate
   b the enzyme.

2.3 Nucleic acids and protein synthesis

Nucleic acids are the information molecules of cells. They fulfil this role throughout the living world, for the code containing the information in nucleic acids, known as the **genetic code**, is a universal one. That means that it is not specific to any one organism or even to a larger group – like mammals, or bacteria – alone. It makes sense in *all* organisms.

There are two types of nucleic acid, referred to as **deoxyribonucleic acid (DNA)**, and **ribonucleic acid (RNA)**. These molecules have roles in the day-to-day control of cells and organisms and in the transmission of the genetic code from generation to generation. First, we need to look at the nucleotides from which nucleic acids are formed.

Structure of nucleotides

Nucleic acids are built up from repeating units called **nucleotides**. A nucleotide consists of three substances combined together (Figure 2.30):

- **a nitrogenous base**, either cytosine (C), guanine (G), adenine (A), thymine (T), or uracil (U)
- **a pentose sugar** (contains five carbon atoms), either ribose or deoxyribose
- **phosphoric acid**.

Nucleotides become condensed together to form huge molecules – the nucleic acids, also known as polynucleotides. A nucleic acid or polynucleotide is a very long, thread-like macromolecule. Alternating sugar and phosphate molecules form the 'backbone' of the polynucleotide, with a nitrogenous base attached to each sugar molecule along the strand (Figure 2.31).

Figure 2.30 The components of nucleotides.

the components:

phosphoric acid

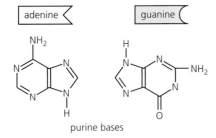

pentose sugars

ribose   deoxyribose

nitrogenous bases

adenine   guanine   thymine   uracil   cytosine

purine bases   pyrimidine bases

condensation to form a nucleotide:

phosphoric acid

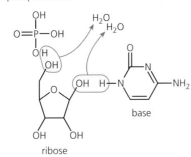

base

ribose

shown diagrammatically as:

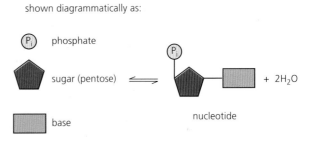

P_i phosphate

sugar (pentose)   nucleotide

base

$+ 2H_2O$

Figure 2.31 How nucleotides make up nucleic acid.

nucleotides are added at this end of the growing polynucleotide

H_2O +

H_2O +

H_2O +

H_2O +

Nucleotides become chemically combined together, phosphate to pentose sugar, by covalent bonds, with a sequence of bases attached to the sugar residues. Up to 5 million nucleotides condense together in this way, forming a polynucleotide (nucleic acid).

The differences between RNA and DNA

RNA molecules are relatively short in length, compared with DNA. In fact, RNAs tend to be between 100 and thousands of nucleotides long, depending on the particular role they have. In all RNA molecules, the nucleotides contain ribose. The other chemical feature that distinguishes RNA is that the bases are cytosine, guanine, adenine and uracil, but never thymine (Figure 2.32). RNA is always a single strand.

In the 'information business' of cells there are three functional types of RNA, known as **messenger RNA (mRNA)**, **transfer RNA (tRNA)**, and **ribosomal RNA**. Their roles are to use information from the nucleus in the construction of proteins by the ribosomes in the cytoplasm.

Figure 2.32 RNA structure.

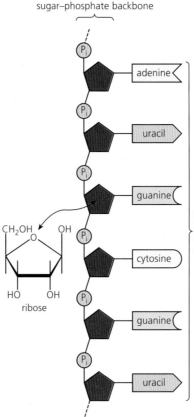

sugar–phosphate backbone

single strand of polynucleotide with ribose sugar and nitrogenous bases: adenine, uracil, guanine and cytosine

DNA molecules occur in the chromosomes and form very long strands, of the order of several million nucleotides in length. RNA molecules are very short, by comparison. In all DNA molecules the nucleotides contain deoxyribose. The bases in DNA are cytosine, guanine, adenine and thymine, but never uracil.

Another distinctive structural feature is that DNA consists of two polynucleotide strands, paired together, and held by hydrogen bonds (page 2). The two strands take the shape of a double helix (Figure 2.33). The pairing of bases is between adenine and thymine, and between cytosine and guanine, because these are the only combination of bases that will fit together within the helix. This pairing, known as **complementary base pairing**, is the key to the way information is held in nucleic acids, and the form in which it can be transferred to RNA (mRNA) to be used in the cytoplasm.

Activity 2.12: A table of the differences between DNA and RNA

A* Extension 2.7: How was the structure of DNA discovered?

Figure 2.33 DNA structure.

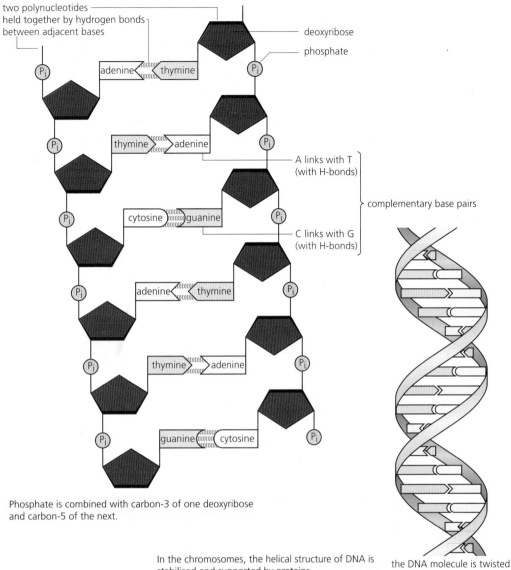

two polynucleotides held together by hydrogen bonds between adjacent bases

deoxyribose

phosphate

adenine — thymine

thymine — adenine

A links with T (with H-bonds)

complementary base pairs

cytosine — guanine

C links with G (with H-bonds)

adenine — thymine

thymine — adenine

guanine — cytosine

Phosphate is combined with carbon-3 of one deoxyribose and carbon-5 of the next.

In the chromosomes, the helical structure of DNA is stabilised and supported by proteins.

the DNA molecule is twisted into a double helix

Replication – how DNA copies itself

In order that a copy of each chromosome can pass into each daughter cell at cell division, the DNA of the chromosomes must first be copied (replicated). As the DNA carries the genetic message, replication must be extremely accurate. Replication is a process that takes place in the interphase nucleus, well before the events of nuclear division (Chapter 3, page 109).

In replication, the DNA double helix must first unwind, and the hydrogen bonds holding the strands together must be broken. This allows the two strands of the helix to separate. A specific enzyme, **helicase**, is involved in these steps and also holds the strands apart while replication occurs.

Both strands of DNA act as templates in replication. New nucleotides with the appropriate complementary bases line up opposite the bases of the exposed strands. Adenine pairs with thymine, cytosine with guanine – a process called **base pairing**. Because of base pairing, the sequence of bases in one strand exactly determines the sequence of bases in the other strand. The two strands are said to be **complementary**. The bases of the two strands fit together only if the sugar molecules they are attached to point in opposite directions. The strands are said to be **antiparallel**.

Hydrogen bonds then form between these complementary bases, holding them in place. Finally the sugar and phosphate groups of adjacent nucleotides of the new strand condense together. This reaction is catalysed by an enzyme called **DNA polymerase**. DNA replication is summarised in Figure 2.34.

Then each pair of strands winds up into a double helix. One strand of each new double helix came from the original chromosome and one is a newly synthesised strand. This arrangement is known as **semi-conservative replication** (Figure 2.35). (If an entirely new double strand had been formed alongside the original, then one DNA double helix would be conserved without unzipping – conservative replication – which is not the case.)

DNA polymerase also has a role in 'proof reading' the new strands. Any 'mistakes' that start to happen (for example, the wrong bases attempting to pair up) are corrected. Each new DNA double helix is an exact copy of the original.

Figure 2.34 DNA replication.

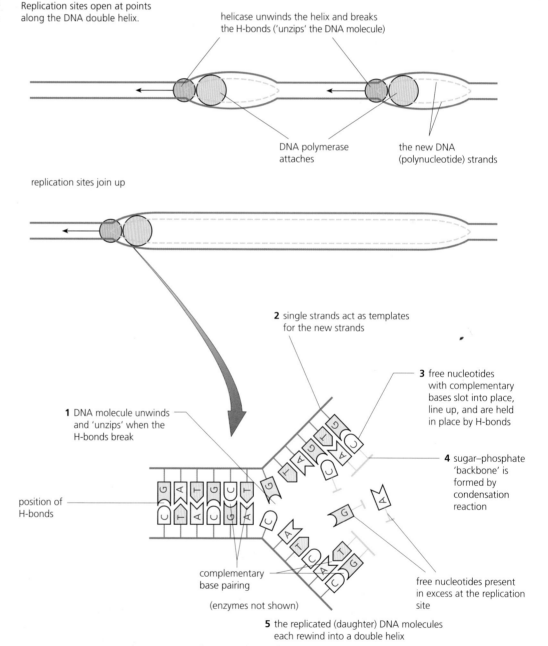

Replication sites open at points along the DNA double helix.

helicase unwinds the helix and breaks the H-bonds ('unzips' the DNA molecule)

DNA polymerase attaches

the new DNA (polynucleotide) strands

replication sites join up

2 single strands act as templates for the new strands

3 free nucleotides with complementary bases slot into place, line up, and are held in place by H-bonds

1 DNA molecule unwinds and 'unzips' when the H-bonds break

4 sugar–phosphate 'backbone' is formed by condensation reaction

position of H-bonds

complementary base pairing

(enzymes not shown)

free nucleotides present in excess at the replication site

5 the replicated (daughter) DNA molecules each rewind into a double helix

The evidence for DNA replication

Experimental confirmation that DNA is replicated semi-conservatively, and not by some other mechanism, came from an experiment by Meselson and Stahl in which a culture of a bacterium (*Escherichia coli*) was grown in a medium (food source) where the available nitrogen contained only the heavy nitrogen isotope, ^{15}N. Consequently, the DNA of the bacterium became entirely 'heavy'.

These bacteria were then transferred to a medium containing the normal (light) isotope, ^{14}N. New DNA manufactured by the cells was now made of ^{14}N. The change in concentration of ^{15}N and ^{14}N in the DNA of succeeding generations was measured. Interestingly, the bacterial cell divisions in a culture of *E. coli* are naturally synchronised; every 60 minutes they all divided again.

The DNA was extracted from samples of the bacteria from each succeeding generation and the DNA in each sample was separated. This was done by placing the sample on top of a salt solution of increasing density, in a centrifuge tube. On being centrifuged, the different DNA molecules were carried down to the level where the salt solution was of the same density. Thus, DNA with 'heavy' nitrogen ended up nearer the base of the tubes, whereas DNA with 'light' nitrogen stayed near the top of the tubes. Figure 2.35 shows the results that were obtained.

DL
www
Activity 2.13: Learning about isotopes

14 Predict the experimental results you would expect if the Meselson–Stahl experiment (Figure 2.35) was carried on for three generations.

1 Meselson and Stahl 'labelled' nucleic acid (i.e. DNA) of the bacterium *Escherichia coli* with 'heavy' nitrogen (^{15}**N**), by culturing in a medium where the only nitrogen available was as $^{15}NH_4{}^{+}$ ions, for several generations of bacteria.

2 When DNA from labelled cells was extracted and centrifuged in a density gradient (of different salt solutions) all the DNA was found to be 'heavy'.

3 In contrast, the DNA extracted from cells of the original culture (before treatment with ^{15}**N**) was 'light'.

4 Then a labelled culture of *E.coli* was switched back to a medium providing unlabelled nitrogen only, i.e. $^{14}NH_4{}^{+}$. Division in the cells was synchronised, and:
- after **one generation** all the DNA was of intermediate density (each of the daughter cells contained (i.e. *conserved*) one of the parental DNA strands containing ^{15}**N** alongside a newly synthesised strand containing DNA made from ^{14}N)
- after **two generations** 50% of the DNA was intermediate and 50% was 'light'. This too agreed with semi-conservative DNA replication, given that labelled DNA was present in only half the cells (one strand per cell).

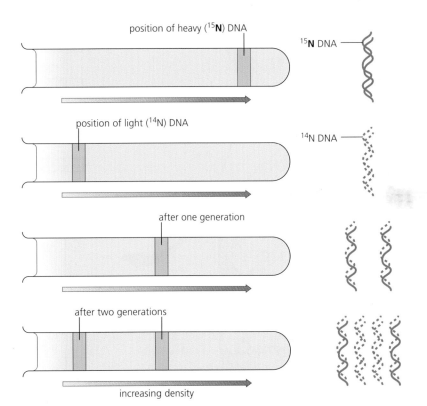

position of heavy (^{15}N) DNA

^{15}**N** DNA

position of light (^{14}N) DNA

^{14}N DNA

after one generation

after two generations

increasing density

Figure 2.35 DNA replication is semi-conservative.

DNA in protein synthesis – the genetic code

In effect, the role of DNA is to instruct the cell to make specific proteins. The huge length of the DNA molecule in a single chromosome codes for a very large number of proteins. Within this extremely long molecule, the relatively short length of DNA that codes for a sequence of amino acids in a polypeptide chain is called a **gene**. Proteins are very variable in size and therefore, so are genes. A very few genes are as short as 75–100 nucleotides long. Most are at least 1000 nucleotides in length, and some are more.

Most proteins contain several hundred amino acids condensed together in a linear series. There are only 20 or so amino acids which are used in protein synthesis; all cell proteins are built from them. The unique properties of each protein lie in:

■ which amino acids are involved in its construction
■ the sequence in which these amino acids are joined.

The DNA code is in the form of a sequence of the four bases, cytosine (C), guanine (G), adenine (A) and thymine (T). This sequence dictates the order in which specific amino acids are to be assembled and combined together. The code lies in the sequence in one of the strands, the reference or **coding strand** (Figure 2.36). The other strand is complementary to the reference strand. The coding strand is always read in the same direction.

The code is a three-letter or **triplet code**, meaning that each sequence of three of the four bases stands for one of the 20 amino acids, and is called a **codon**. With a four-letter alphabet (C,G,A,T) there are 64 possible different triplet combinations (4 × 4 × 4). In other words, the genetic code has many more codons than there are amino acids. In fact most amino acids have two or three similar codons that code for them. Also, some of the codons represent the 'punctuations' of the code – for example, there are 'start' and 'stop' triplets (Figure 2.37).

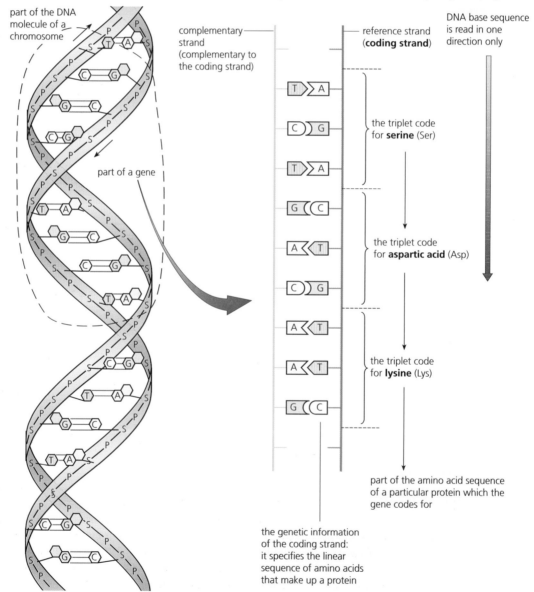

Figure 2.36 Part of a gene and how its DNA codes for amino acids.

Figure 2.37 The genetic code – a universal code.

The 20 amino acids used in protein synthesis

Amino acids	Abreviations
alanine	Ala
arginine	Arg
asparagine	Asn
aspartic acid	Asp
cysteine	Cys
glutamine	Gln
glutamic acid	Glu
glycine	Gly
histidine	His
isoleucine	Ile
leucine	Leu
lysine	Lys
methionine	Met
phenylalanine	Phe
proline	Pro
serine	Ser
threonine	Thr
tryptophan	Trp
tyrosine	Tyr
valine	Val

The genetic code in circular form

The codons are those of messenger RNA (where uracil, U, replaces thymine, T)

Read the code from the centre of the circle outwards along a radius. For example, serine is coded by UCU, UCC, UCA or UCG, or by AGU or AGC.

In addition, some codons stand for 'stop', signalling the end of a peptide or protein chain.

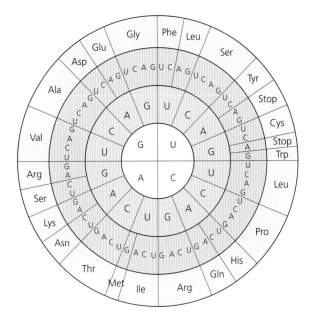

15 The sequence of bases in a sample of mRNA was found to be:
GGU,AAU,CCU,UUU,GUU,ACU,CAU,UGU.
   a What sequence of amino acids does this code for?
   b What was the sequence of bases in the coding strand of DNA from which this mRNA was transcribed?

The stages of protein synthesis

There are three stages in the process by which the information of the gene is used to determine how the protein molecule is constructed.

Stage 1 occurs in the nucleus where a complementary copy of the code is made by the building of a molecule of **messenger RNA (mRNA)**. This process is called **transcription** and the enzyme is **RNA polymerase**. One strand of the DNA, the coding strand, is used as the **template**. Once the mRNA strand is formed it leaves the nucleus through pores in the nuclear membrane (Figure 2.38) and passes to tiny structures in the cytoplasm called **ribosomes** where the information can be 'read' and is used.

In **Stage 2**, the amino acids are activated for protein synthesis by combining with short lengths of a different sort of RNA, called **transfer RNA (tRNA)**. This activation occurs in the cytoplasm. It is the tRNA that translates a three-base sequence into an amino acid sequence.

All the tRNAs take the shape of a clover-leaf, but there is a different tRNA for each of the 20 amino acids involved in protein synthesis. At one end of each tRNA molecule is a site where a particular amino acid can be joined (Figure 2.39). At the other end, there is a sequence of three bases called an **anticodon**. This anticodon is complementary to the codon of mRNA that codes for the specific amino acid.

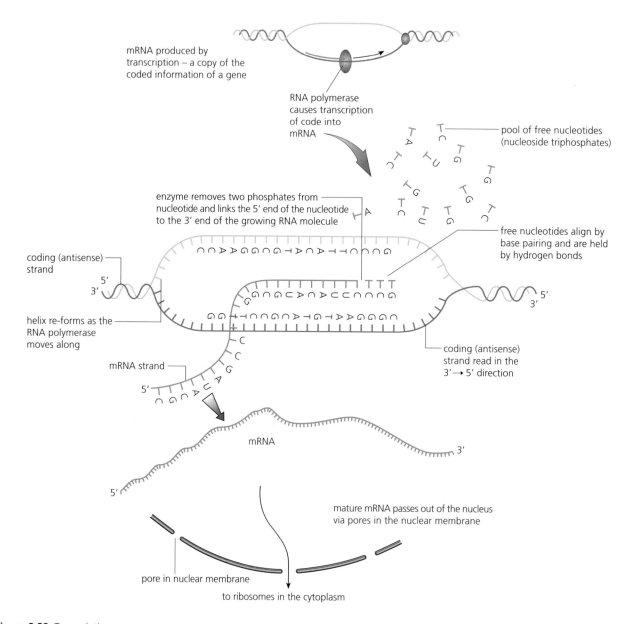

Figure 2.38 Transcription.

The amino acid is attached to its tRNA by an enzyme. These enzymes are specific to the particular amino acids (and types of tRNA) to be used in protein synthesis. The specificity of the enzymes is a way of ensuring the correct amino acids are used in the right sequence.

In **Stage 3**, a protein chain is assembled, one amino acid residue at a time. The process is called **translation** (Figure 2.40). Tiny organelles called ribosomes move along the messenger RNA 'reading' the codons from a 'start' codon. In the ribosome, complementary anticodons on the amino acid–tRNAs slot into place and are temporarily held in position by **hydrogen bonds**. While held there, the amino acids of neighbouring amino acid–tRNAs are joined by peptide linkages. This frees the first tRNA which moves back into the cytoplasm for re-use. Once this is done, the ribosome moves on to the next mRNA codon. The process continues until a 'stop' codon occurs.

A* Extension 2.8: Transfer RNA – a small molecule that plays an important role in protein synthesis

Activity 2.14: Wider reading – 'Exploring proteins'

16 What different forms of RNA are involved in *transcription* and *translation*, and what are the roles of each?

Each amino acid is linked to a specific transfer RNA (tRNA) before it can be used in protein synthesis. This is the process of amino acid activation. It takes place in the cytoplasm.

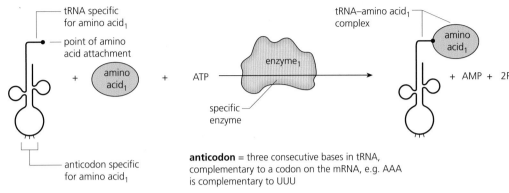

anticodon = three consecutive bases in tRNA, complementary to a codon on the mRNA, e.g. AAA is complementary to UUU

Figure 2.39 Amino acid activation.

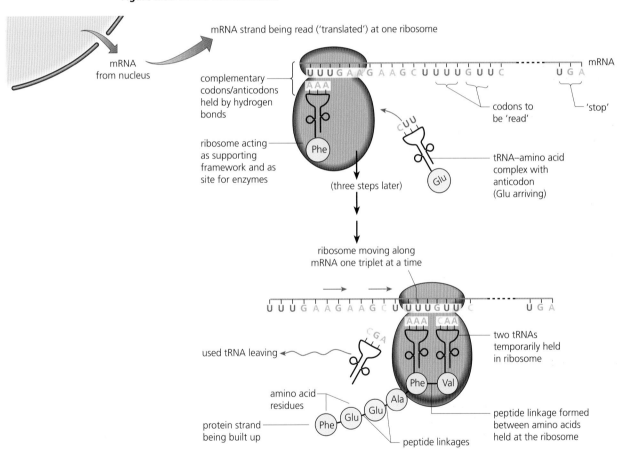

Figure 2.40 Translation.

DNA can change

We have seen that a gene is a sequence of nucleotide pairs which codes for a sequence of amino acids. Normally, the sequence of nucleotides in DNA is maintained without changing, but very occasionally it does change. If there is a change, we say a **mutation** has occurred. A mutation is an unpredictable change in the genetic make-up of a cell. At certain times in the cell cycle and under particular external conditions, mutations are more likely than at other times. One such occasion is when the DNA molecule is replicating.

Actually, two types of mutation are possible – chromosome mutations and gene mutations.

Figure 2.41 Sickle cell anaemia: an example of a gene mutation.

Anaemia is a disease typically due to a deficiency in healthy red cells in the blood.

Haemoglobin occurs in red cells – each contains about 280 million molecules of haemoglobin. A molecule consists of two α-haemoglobin and two β-haemoglobin subunits, interlocked to form a compact molecule.

The **mutation** that produces sickle cell haemoglobin (**Hg^S**) is in the gene for β-haemoglobin. It results from the substitution of a single base in the sequence of bases that make up all the codons for β-haemoglobin.

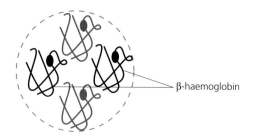

β-haemoglobin

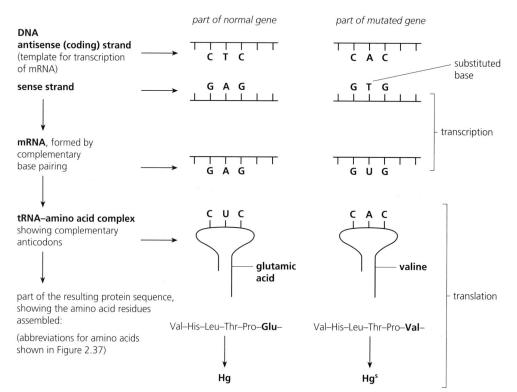

DNA
antisense (coding) strand (template for transcription of mRNA)

part of normal gene   *part of mutated gene*

C T C   C A C

substituted base

sense strand

G A G   G T G

transcription

mRNA, formed by complementary base pairing

G A G   G U G

tRNA–amino acid complex showing complementary anticodons

C U C   C A C

glutamic acid   valine

translation

part of the resulting protein sequence, showing the amino acid residues assembled:

(abbreviations for amino acids shown in Figure 2.37)

Val–His–Leu–Thr–Pro–**Glu**–   Val–His–Leu–Thr–Pro–**Val**–

Hg   **Hg^s**

drawing based on a photomicrograph of a blood smear, showing blood of a patient with sickle cells present among healthy red cells

phenotypic appearance of **HgHg** red cells and sickle cells (**Hg^{s}Hg**)

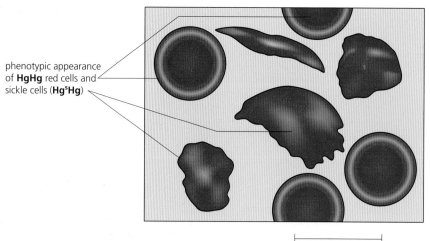

10 μm

Chromosome mutations

A chromosome mutation occurs when an abrupt change in the number or the sequence of genes occurs. It may be brought about in a number of different ways. For example, additional sets of chromosomes may be the result of an error during the nuclear divisions of gamete formation. The cultivated potato, which has double the number of chromosomes that the smaller, wild potato has, is a case in point. Alternatively, there may be an alteration to just part of the chromosome set. For example, people with Down's syndrome have an extra chromosome 21, giving them a total of 47 chromosomes (page 101).

Gene mutations

A gene mutation involves a change in the sequence of bases of a particular gene. We have already noted that the enzyme machinery that brings about the building of a complementary DNA strand also 'proof reads' and corrects most errors. However, gene mutations can and do occur spontaneously during this step. Certain chemicals may also cause change to the DNA sequence of bases. So do some forms of radiation, such as X-rays. Factors that increase the chances of a mutation are called **mutagens**.

For example, it is a gene mutation that is responsible for the condition known as sickle cell anaemia. Here, the gene that codes for the amino acid sequence of a part of the respiratory pigment haemoglobin, found in our red cells, is prone to a mutation in one base pair. This causes valine to appear at that point, instead of glutamic acid (Figure 2.41). The resulting, abnormal haemoglobin tends to clump together and form long fibres that distort the red cells into sickle shapes. In this condition they cannot transport oxygen and the cells may block smaller capillaries.

Another example is the inherited disorder known as cystic fibrosis, and this is discussed below (page 90).

DL
www
Activity 2.15: Wider reading – 'A chromosomal rearrangement'

2.4 Introducing genetics

Genetics is the science of heredity. It is the study of how variation arises and how the characteristics of individuals are passed from generation to generation. We have noted that the nucleus, in addition to being the organising centre of a cell, is the location of the hereditary material which is passed from generation to generation during reproduction. The information the nucleus holds on its chromosomes exists within DNA; a single DNA molecule runs the length of each chromosome, and is effectively a linear series of **genes**.

We can define 'gene' in different ways. For example, a gene is:

■ a specific region of a chromosome which is capable of determining the development of a specific characteristic of an organism
■ a specific length of the DNA double helix, hundreds or (more typically) thousands of base pairs long, which codes for a sequence of amino acids in a polypeptide chain
■ a unit of inheritance.

A particular gene always occurs on the same chromosome in the same position. The position of a gene is called its **locus** (plural loci). As we will see shortly, each gene may have two or more forms, called **alleles**. The word allele just means 'alternative forms'.

Introducing chromosomes

There are four characteristic features about the chromosomes of organisms which it is helpful to note at the outset.

1 **The number of chromosomes per species is fixed.** The number of chromosomes in the cells of different species varies, but in any one species the number of chromosomes per cell is normally constant. For example, the mouse has 40 chromosomes per cell, the onion has 16, humans have 46, and the meadow buttercup 14. These are the chromosome numbers for the species. Note that these are all even numbers.

2 **Chromosomes occur in pairs.** The chromosomes of a cell occur in pairs, called **homologous pairs**. Homologous means 'similar in structure'. One set of chromosomes came originally from one parent and a second set from the other. So, for example, the human has 46 chromosomes, 23 coming originally from each parent. This is why they occur in homologous pairs.

You can see this from the photomicrographs in Figure 2.42. Here, human chromosomes are shown at a stage of nuclear division, and also cut out and arranged in descending order of size. Traditionally, homologous pairs are numbered in this way in order to be able to identify them.

Figure 2.42
Chromosomes as homologous pairs.

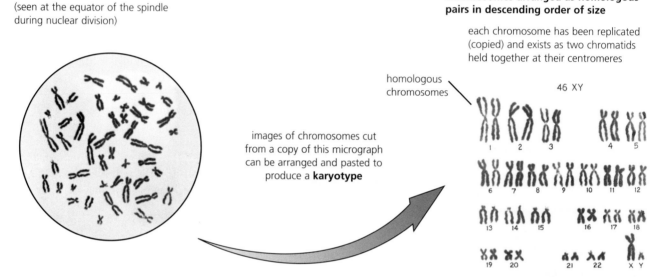

human chromosomes of a male
(seen at the equator of the spindle during nuclear division)

chromosomes arranged as homologous pairs in descending order of size

each chromosome has been replicated (copied) and exists as two chromatids held together at their centromeres

homologous chromosomes

46 XY

images of chromosomes cut from a copy of this micrograph can be arranged and pasted to produce a **karyotype**

3 **The shape of a chromosome is characteristic.** Chromosomes are long thin structures of a particular, fixed length. Somewhere along the length of the chromosome occurs a characteristically narrow region called the centromere. A centromere may occur anywhere along the chromosome, but it is always in the same position on any given chromosome. The position of its centromere, as well as the length of a chromosome, allows us to identify it in photomicrographs.

4 **Chromosomes copy themselves.** Between nuclear divisions, each chromosome makes a copy of itself. It is said to replicate. The process of replication has already been discussed (page 73). DNA replication is one particular phase (S) of the cell cycle (Chapter 3, page 108).

Introducing variation

Genetics is the study of variation and its inheritance.

What is meant by variation in this context?
Individuals of a species are strikingly similar, which is how we may identify them, whether humans, buttercups or houseflies, for example. Individuals also show numerous differences, although we may have to look carefully for them initially, in members of species other than our own. Certainly, within families there are remarkable similarities between parents and their offspring, but no two members of a family are identical, apart from identical twins.

The variation we observe in living things may be due to genetics, or to an effect of the environment on the individual, or both.

■ Genetic differences are controlled by genes – for example, the sex of an individual depends on which sex chromosomes it possesses.
■ Other variations between individuals are due to the environment they experience, such as effects of an illness contracted, or deficiency of nutrients.
■ Others may be due to both genetics and environment, such as our body height and weight.

Another important point about variations is that they are of two types.

- **Discontinuous variations** are ones in which the characteristic concerned is one of two or more discrete types with no intermediate forms. Examples include the garden pea plant (tall or dwarf), and human ABO blood grouping (A, B, AB or O group). These are genetically determined.
- **Continuous variations** show a continuous distribution of values. Height in humans is a good example. Continuous variation may be genetically determined, or it may be due to environmental and genetic factors working together.

Figure 2.43
Discontinuous and continuous variables.

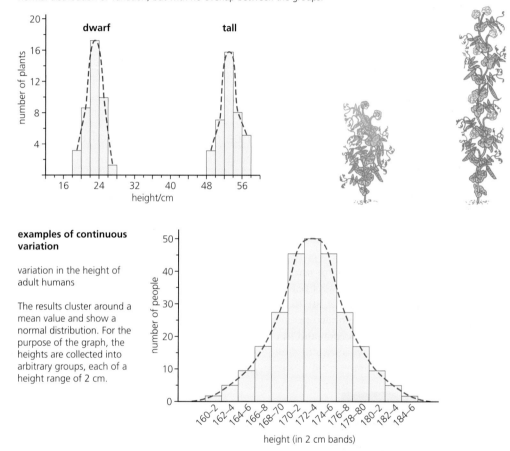

examples of discontinuous variation

The heights of these plants fall into two discrete groups, in both of which there is a normal distribution of variation, but with no overlap between the groups.

examples of continuous variation

variation in the height of adult humans

The results cluster around a mean value and show a normal distribution. For the purpose of the graph, the heights are collected into arbitrary groups, each of a height range of 2 cm.

Inheriting genes in sexual reproduction

In sexual reproduction, **gametes** (sex cells – sperms and ova) are formed by a special reduction division of the nucleus. These gametes are **haploid**, meaning each has one set of chromosomes (only one of each homologous pair); it is the basic set. So gametes contain only one copy of each gene. At fertilisation, male and female gametes fuse to form a **zygote**. Consequently, the zygote is a **diploid** cell, with two sets of chromosomes (homologous pairs), one from each parent. Thus there are two copies of each gene (alleles) in the new individual.

The alleles that an organism has present in every cell make up the **genotype** of that organism. So the genotype is the genetic constitution of an organism. A genotype in which the two alleles of a gene are the same is said to be **homozygous** for that gene. If the alleles are different, the organism is **heterozygous** for that gene.

The whole genotype may interact with environmental factors. The outcome is the **phenotype**. The phenotype is the way in which the genotype of the organism is expressed – that is, the appearance of the organism (Table 2.5).

17 Explain why:
a chromosomes occur in homologous pairs in diploid cells
b it is essential that nuclear division is a precise process.

Table 2.5 Essential genetic terms.

Term	Definition
genotype	the genetic constitution of an organism
phenotype	the characteristics displayed by the organism – the way in which the genotype is expressed (often, the appearance of an organism)
gene	the basic unit of inheritance by which inherited characteristics are transferred from parents to offspring, consisting of a length of DNA on a chromosome
alleles	alternative forms of a gene, occupying a specific position (locus) on a chromosome
dominant allele	an allele that affects the phenotype of the organism whether present in the heterozygous or homozygous condition
recessive allele	an allele that affects the phenotype of the organism only when the dominant allele is absent (i.e. in homozygous recessive individuals)
homozygous	a diploid organism that has inherited the same allele (for any particular gene) from both parents
heterozygous	a diploid organism that has inherited different alleles from each parent

We can now apply these genetic definitions as we consider a breeding experiment.

Pattern of inheritance of a single pair of contrasting characteristics

The mechanism of inheritance was first successfully investigated before chromosomes had been observed – in the mid-nineteenth century, by Gregor Mendel. His experiments involved the inheritance of contrasting characteristics of the garden pea plant, *Pisum sativum*, such as the height of the stem. The stem may be either 'tall' (say about 48 cm), or 'dwarf' (about 12 cm). We now know that this characteristic is controlled by a single gene with two alleles. A breeding experiment in which the inheritance of the alleles of a single gene is investigated is known as a **monohybrid cross**.

The plants used as the **parent generation (P)** in this cross were one tall and one dwarf plant, and they were pure-breeding plants. 'Pure-breeding' plants, when crossed among themselves always give rise to offspring like the parents. The genetics term for pure-breeding is **homozygous**.

The garden pea plant naturally self-pollinates (and therefore fertilises itself), even though the flowers open and insects such as the honey bee visit for pollen and nectar, and inadvertently

three steps in the cross-pollination of the pea plant

1 Using an immature flower, the keel is cut open to expose the immature stamens.

2 Pollen from a mature pea flower is introduced to the mature stigma.

hairs of paintbrush deliver pollen from another flower

stigma
anther
filament
} stamen

stamens (anthers and stalks of the filaments) are cut off and thrown away

filaments form tube around the ovary

stigma (pollen introduced here)
style
ovary surrounded by remains of the filaments

3 Next, the arrival of stray pollen is prevented by covering the flower with a muslin bag.

standard
wings
keels

Flowers of the pea family have five petals: a rear **standard**, two lateral **wings**, and two **keels** arranged like a boat, surrounding the male (stamens) and female (stigma, style and ovary) parts of the flower.

Insects, such as bees, part the keels on landing, expose the stamens and stigma, and may pollinate the flower.

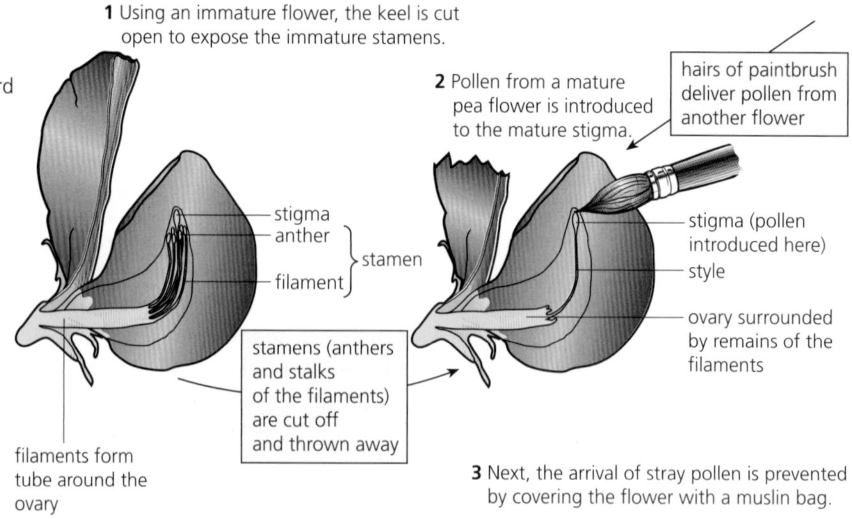

The garden pea plant is naturally self-pollinated, even though it is visited by insects.

Figure 2.44 Cross pollination of the garden pea plant.

transport pollen between flowers. So, in order to carry out a genetic 'cross' between different garden pea plants, self-pollination had to be prevented. To do this, flowers had their male parts (the stamens) cut out while still immature (before pollen had been formed). Then, pollen was introduced from a flower on a plant with the contrasting characteristic (that is, pollen from a flower of a tall plant was introduced onto the female part – the stigma – of a dwarf plant, and vice versa), as shown in Figure 2.44.

The offspring from this cross were a large number of peas, which, when planted, all grew into tall plants. The offspring are known as the **first filial generation**, shortened to **F₁**. Plants of the F_1 generation were then allowed to self-pollinate (and so self-fertilise) to form the **second filial generation** or **F₂**. The offspring, a large number of pea seeds, when planted, grow into a mixture of tall and dwarf plants in the ratio of 3 tall to 1 dwarf (Figure 2.45).

How is this brought about? What is the significance of this ratio?

Figure 2.45 The monohybrid cross in summary.

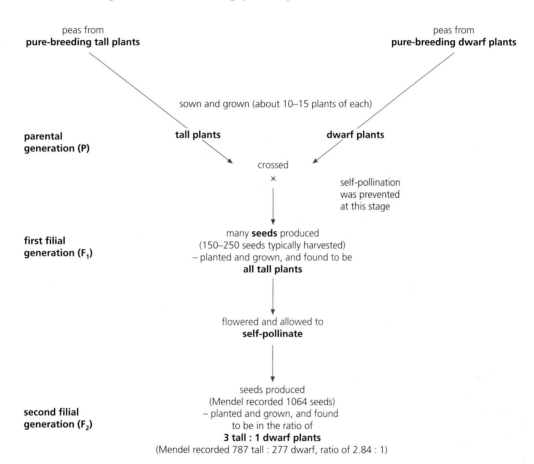

Interpreting the monohybrid cross

We can understand the outcome of the monohybrid cross in terms of the behaviour of genes and alleles in gamete formation, fertilisation and in development of the offspring. First we choose the symbol **T** to represent the allele for tall and the symbol **t** for the allele for dwarf.

The parents were homozygous for height alleles, meaning that the parent of phenotype 'tall' had a genotype **TT** and the parent of phenotype 'dwarf' had a genotype **tt**. The parents each produce only one type of gamete (involving the reduction division; the tall parent produced gametes containing a single allele for 'tall', **T**, and the dwarf parent produced gametes containing a single allele for 'dwarf', **t**).

Fertilisation between these gametes produced offspring with a genotype that is heterozygous for height alleles, **Tt**. The phenotype of the F_1 generation was 'tall', so we see that, in the heterozygous nucleus, the 'tall' allele is **dominant** in the presence of the 'dwarf' allele, which is **recessive**. Geneticists say that the dominant allele is **expressed** in the phenotype. (Note that the recessive allele is not lost or destroyed, but rather is temporarily inactive or ineffective.)

The F_1 generation are all heterozygous tall, **Tt**, and when they form gametes, half will carry the **T** allele and half the **t** allele. So, each heterozygous parent produces two types of gamete. Fertilisation involves the random fusion of male and female gametes. So a **T** male gamete may fuse with a **T** or a **t** egg cell, and a **t** male gamete may fuse with a **T** or a **t** egg cell. The outcome is easily shown in a matrix called a **Punnett grid** (in Figure 20.46), named after the biologist who first used it.

As a consequence, *provided a large number of offspring are formed* from the cross, about a quarter of the offspring are homozygous tall (**TT**), half are heterozygous tall (**Tt**) and a quarter are homozygous dwarf (**tt**). Meanwhile, the ratio of the phenotypes is 3 tall to 1 dwarf. This ratio is typical of a monohybrid cross where one allele is dominant and one allele is recessive.

A genetic cross is represented as a **genetic diagram** (Figure 2.47) which shows the genotypes of parents, gametes and offspring, and explains the ratios to be anticipated in the offspring.

Another outcome of this breeding experiment is that we can confirm that the characteristics of an organism are controlled by pairs of alleles which separate in equal numbers into different gametes at gamete formation. This is known as the principle or **law of segregation**.

In the pea plant, 'height' is controlled by a single gene.

In this cross, a pea plant homozygous for 'tall' was crossed with a pea plant homozygous for 'dwarf'. The offspring were allowed to self-pollinate (and to self-fertilise) to produce the second generation.

The garden pea plant has seven pairs of chromosomes per nucleus (and therefore seven chromosomes in its gametes), but only one pair is represented here, for clarity.

The offspring of self-fertilising ('selfing') heterozygous tall pea plants were in the ratio 3 tall : 1 dwarf.

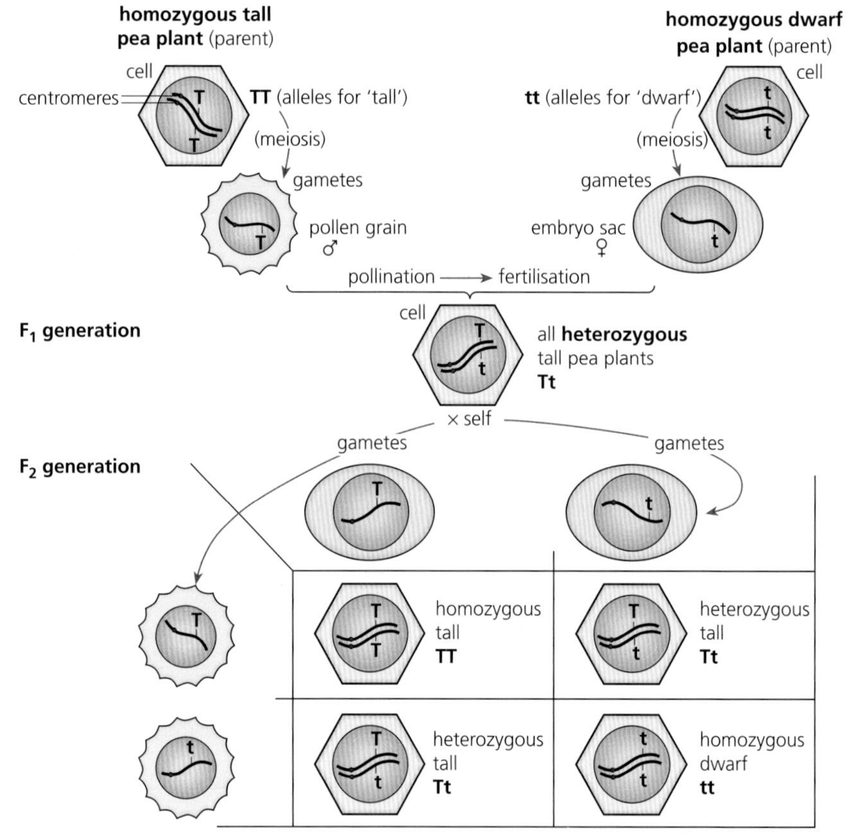

Figure 2.46 The behaviour of alleles in the monohybrid cross.

A* Extension 2.9: The test cross – telling 'tall' pea plants apart

A* Extension 2.10: How did Mendel make his great 'breakthrough'?

18 Another example of monohybrid inheritance in a garden pea plant is the inheritance of the contrasting characteristic 'round' or' wrinkled' seed (symbols **R** and **r**). The outcome of the F_2 generation was a ratio of 3 'round' to 1 'wrinkled' seeds. Construct a genetic diagram of the F_1 and F_2 crosses to explain this outcome.

Figure 2.47 Genetic diagram of the monohybrid cross.

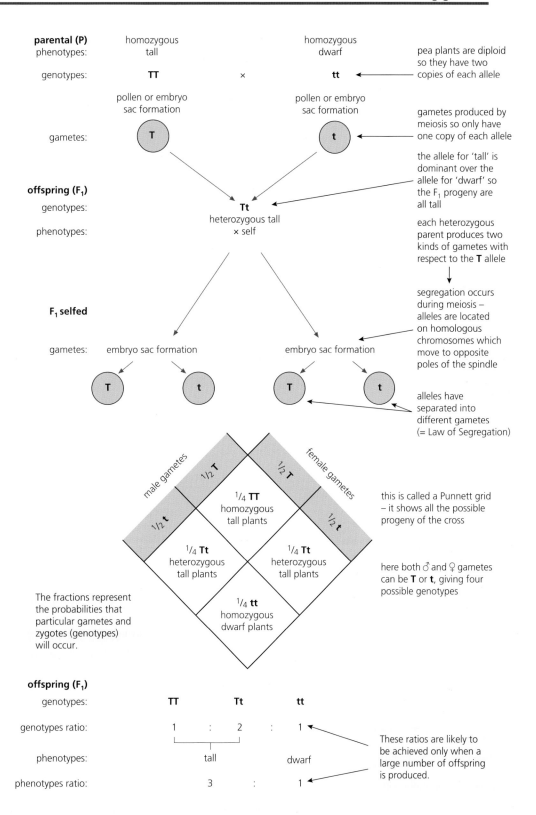

parental (P)
phenotypes: homozygous tall   homozygous dwarf

genotypes: **TT**   ×   **tt**

pea plants are diploid so they have two copies of each allele

pollen or embryo sac formation   pollen or embryo sac formation

gametes: T   t

gametes produced by meiosis so only have one copy of each allele

offspring (F₁)
genotypes: **Tt**
phenotypes: heterozygous tall
                × self

the allele for 'tall' is dominant over the allele for 'dwarf' so the F₁ progeny are all tall

each heterozygous parent produces two kinds of gametes with respect to the **T** allele

F₁ selfed

gametes: embryo sac formation   embryo sac formation

                T   t   T   t

segregation occurs during meiosis – alleles are located on homologous chromosomes which move to opposite poles of the spindle

alleles have separated into different gametes (= Law of Segregation)

male gametes   ½ **T**   ½ **T**   female gametes

½ **t**

¼ **TT** homozygous tall plants

¼ **Tt** heterozygous tall plants

¼ **Tt** heterozygous tall plants

½ **t**

¼ **tt** homozygous dwarf plants

this is called a Punnett grid – it shows all the possible progeny of the cross

here both ♂ and ♀ gametes can be **T** or **t**, giving four possible genotypes

The fractions represent the probabilities that particular gametes and zygotes (genotypes) will occur.

offspring (F₁)
genotypes: **TT**   **Tt**   **tt**

genotypes ratio: 1   :   2   :   1

phenotypes: tall   dwarf

phenotypes ratio: 3   :   1

These ratios are likely to be achieved only when a large number of offspring is produced.

Human inheritance investigated by pedigree chart

Studying human inheritance by experimental crosses (with selected parents, sibling crosses, and the production of large numbers of progeny) is out of the question. Instead, we may investigate the pattern of inheritance of a particular characteristic by researching a family **pedigree**, where appropriate records of the ancestors exist. A human pedigree chart uses a set of rules, outlined in Figure 2.48.

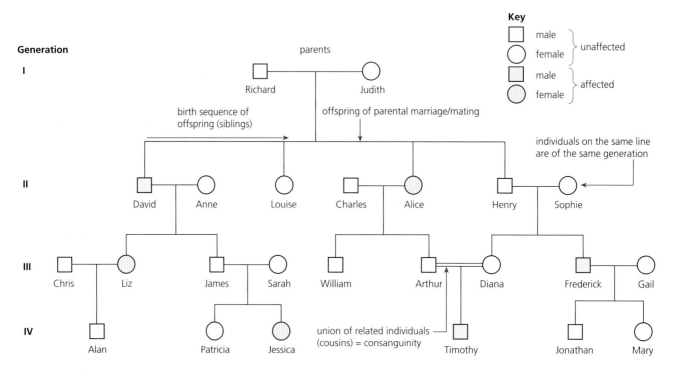

Figure 2.48 An example of a human pedigree chart.

We can use a pedigree chart to detect conditions due to both dominant and recessive alleles. In the case of a characteristic due to a dominant allele, the characteristic tends to occur in one or more members of the family in every generation. On the other hand, a recessive characteristic is seen infrequently, skipping many generations.

An example of a condition caused by a recessive allele is albinism, a rare inherited condition of humans (and other mammals) in which the individual has a block in the biochemical pathway by which the pigment melanin is formed. Albinos have white hair, very light coloured skin and pink eyes. Albinism shows a pattern of recessive monohybrid inheritance in humans (Figure 2.49).

The inheritance of genetic disease

Genetic diseases are heritable conditions that are caused by a specific defect in a gene or genes. Most arise from a mutation involving a single gene. The mutant allele that causes the disease is commonly recessive, and in these cases a person must be homozygous for the mutant gene for the condition to be expressed. However, people with a single mutant allele are 'carriers' of that genetic disease. Quite surprising numbers of us are carriers of one or more conditions.

Genetic diseases generally afflict about 1–2% of the human population. Common genetic diseases include sickle cell anaemia (page 80), Duchenne muscular dystrophy, severe combined immunodeficiency disease (SCID), familial hypercholesterolaemia, and haemophilia, together with the two genetic diseases described next.

Albino people must be homozygous for the recessive albino allele (**pp**). People with normal skin pigmentation may be homozygous normal (**PP**) or carriers (**Pp**).

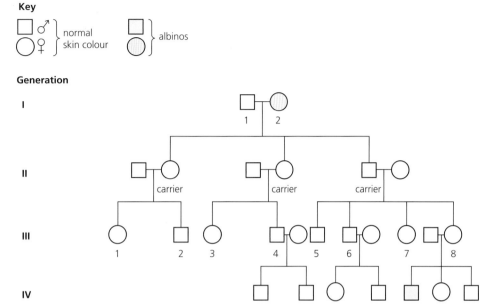

This is a typical family tree for inheritance of a characteristic controlled by a recessive allele.

Figure 2.49 Pedigree chart of a family with albino members.

Other examples of genetic disease

Thalassaemia is caused by recessive alleles of a gene found on chromosome 11, and results from one of several mutations that cause reduced or imperfect synthesis of haemoglobin, most typically in the β chains (see haemoglobin in Figure 2.41, page 80). Sufferers produce normal haemoglobin during fetal development in the uterus, but after birth thalassaemia develops. Sadly, this condition is normally fatal while the patient is still an infant.

The pedigree chart of a family of thalassaemia carriers will have a similar pattern to that shown for albinism. The incidence of this condition is rare among the UK population, but quite high (1 in 130 of the population) among certain communities of south-eastern Europe.

Figure 2.50 Human pedigree chart showing the recessive monohybrid inheritance of cystic fibrosis.

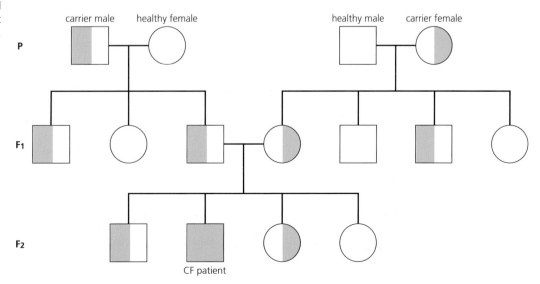

Cystic fibrosis, on the other hand, is the most common genetic disorder in the UK population (one person in 25 is a carrier). Figure 2.50 is a pedigree chart showing the inheritance of cystic fibrosis. It is due to a mutation of a single gene on chromosome 7, and it affects the epithelial cells of the body. The CF gene, 180 000 base pairs long, codes for a protein known as CFTR which functions as an ion pump (Figure 2.51). The pump transports chloride ions across membranes and water follows the ions, so epithelia are kept smooth and moist.

CFTR protein *in situ* in plasma membrane

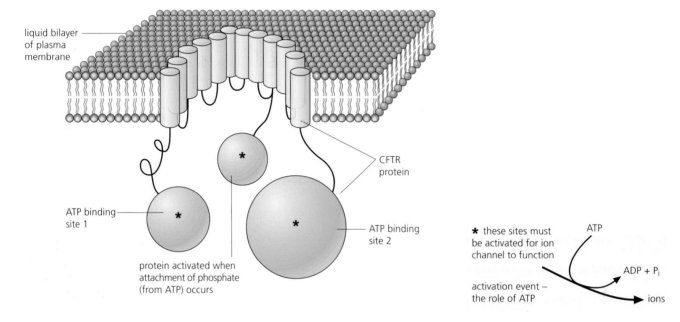

How CFTR regulates water content in mucus

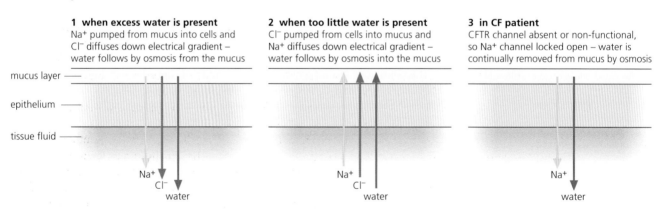

Figure 2.51 The CFTR protein – a channel protein.

In cystic fibrosis, the most common mutation involves a deletion of only three of the gene's nucleotides, and results in the loss of an amino acid (phenylalanine) at one location along a protein built from almost 1500 amino acids in total. The mutated gene codes for no protein or for a faulty protein. As a result, sufferers have epithelia that remain dry, and there is a build-up of thick, sticky mucus. The effects are felt:

- in the pancreas – here, secretion of digestive juices by the gland cells in the pancreas is interrupted by blocked ducts
- in the sweat glands, where salty sweat formed – a feature exploited in diagnosis
- in the lungs, which become blocked by mucus and are prone to infection – this effect is most quickly life-threatening.

male urinogenital system, in section

female urinogenital system, in section

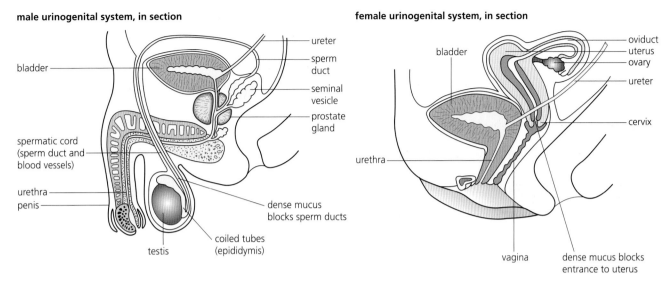

Figure 2.52 Infertility in cystic fibrosis patients.

Activity 2.16: Wider reading – 'Genetic testing: some ethical issues'

A* Extension 2.11: Cystic fibrosis and the flow of pancreatic juice

In adult patients, the membranes of epithelial cells in their reproductive organs are affected too. The mucus that is naturally present in the female cervix, in CF patients becomes a dense plug in the vagina, restricting entry of sperms into the uterus. In males, dense mucus in the sperm ducts blocks these tubes, significantly reducing the sperm count when semen is ejaculated. Infertility frequently results (Figure 2.52).

Introducing gene therapy

Very many organisms have been genetically modified by humans, most of them by means of artificial selection. The majority of the animals and plants used in agriculture, horticulture, transport and leisure pursuits have been bred from wild animals and plants in this way. The effect of this selection process is the relatively speedy and deliberate genetic change of local populations of plants or animals into new varieties, useful to humans.

Today a new type of genetic modification is also in use, known as **genetic engineering**. Genes from one organism are transferred to the set of genes (the genome) of another unrelated organism. The process is also known as **recombinant DNA technology**. Gene technology has important applications in biotechnology, medicinal drug production (in the pharmaceuticals industry), agriculture and horticulture. Genetic engineering generates many potential benefits for humans, but there are potential hazards, too. The economic advantages may be out-numbered by environmental and ethical drawbacks or dangers. The issues require balanced and informed judgements.

One important application is **gene therapy**, the use of recombinant DNA technology to overcome genetic disease, where this is thought safe and ethically sound. Gene therapy is a very recent, and is a highly experimental science. One approach is to supply the missing gene to body cells in such a way that it remains permanently fully functional. Sometimes, all that can be achieved initially is to supply the missing gene on a temporary basis (or even just the gene product) and then periodically re-supply it. These solutions are known as **somatic therapy** (Figure 2.53).

On the other hand, it is not considered safe or ethical to attempt to tamper with germ cells (cells which give rise to gametes, located in the testes and ovaries). This banned approach is called **germ-line therapy**.

What reasons can you think of to account for the ban on germ-line therapy?

The cystic fibrosis gene codes for a membrane protein that occurs widely in body cells, and pumps ions (e.g. Cl⁻) across cell membranes.

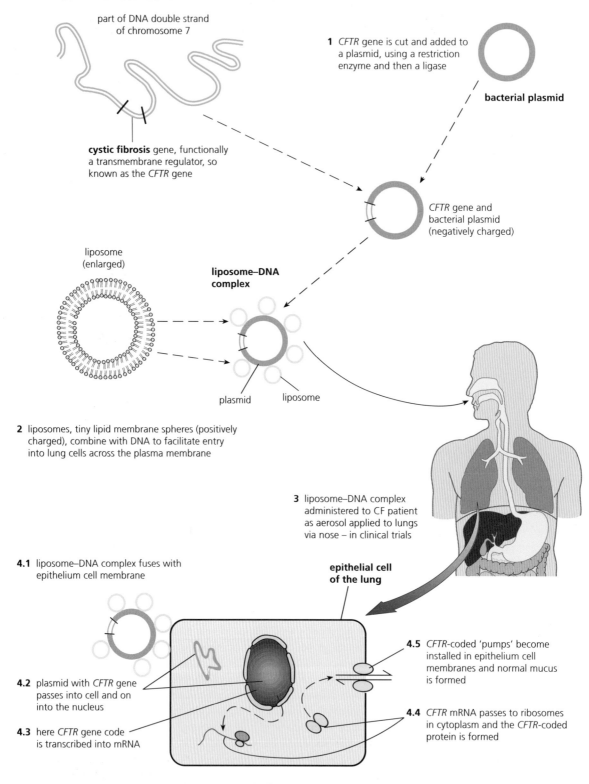

part of DNA double strand of chromosome 7

1 *CFTR* gene is cut and added to a plasmid, using a restriction enzyme and then a ligase

bacterial plasmid

cystic fibrosis gene, functionally a transmembrane regulator, so known as the *CFTR* gene

CFTR gene and bacterial plasmid (negatively charged)

liposome (enlarged)

liposome–DNA complex

plasmid   liposome

2 liposomes, tiny lipid membrane spheres (positively charged), combine with DNA to facilitate entry into lung cells across the plasma membrane

3 liposome–DNA complex administered to CF patient as aerosol applied to lungs via nose – in clinical trials

4.1 liposome–DNA complex fuses with epithelium cell membrane

epithelial cell of the lung

4.5 *CFTR*-coded 'pumps' become installed in epithelium cell membranes and normal mucus is formed

4.2 plasmid with *CFTR* gene passes into cell and on into the nucleus

4.4 *CFTR* mRNA passes to ribosomes in cytoplasm and the *CFTR*-coded protein is formed

4.3 here *CFTR* gene code is transcribed into mRNA

In recent clinical trials some 20% of epithelium cells of CF patients were *temporarily* modified (i.e. accepted the *CFTR* gene), but the effects were relatively short-lived. This is because our epithelium cells are continually replaced at a steady rate, and in CF patients the genetically engineered cells are replaced with cells without *CFTR*-coded pumps. Patients would require periodic treatment with the liposome–DNA complex aerosol to maintain the effect permanently.

Figure 2.53 Somatic gene therapy for cystic fibrosis – supplying the healthy gene to the lungs.

1 DNA extraction → cut with restriction enzyme

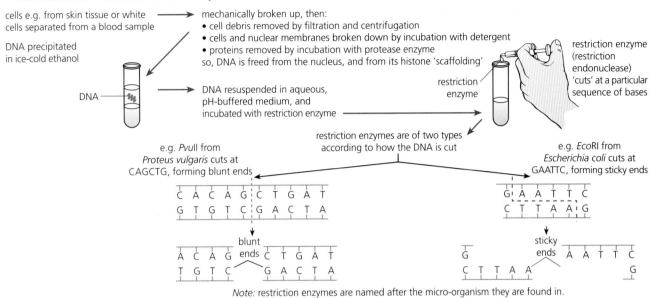

cells e.g. from skin tissue or white cells separated from a blood sample

DNA precipitated in ice-cold ethanol

mechanically broken up, then:
- cell debris removed by filtration and centrifugation
- cells and nuclear membranes broken down by incubation with detergent
- proteins removed by incubation with protease enzyme

so, DNA is freed from the nucleus, and from its histone 'scaffolding'

DNA

restriction enzyme

DNA resuspended in aqueous, pH-buffered medium, and incubated with restriction enzyme

restriction enzyme (restriction endonuclease) 'cuts' at a particular sequence of bases

restriction enzymes are of two types according to how the DNA is cut

e.g. *Pvu*II from *Proteus vulgaris* cuts at CAGCTG, forming blunt ends

C A C A G C T G A T
G T G T C G A C T A

blunt ends

A C A G   C T G A T
T G T C   G A C T A

e.g. *Eco*RI from *Escherichia coli* cuts at GAATTC, forming sticky ends

G A A T T C
C T T A A G

sticky ends

G   A A T T C
C T T A A   G

Note: restriction enzymes are named after the micro-organism they are found in.

2 DNA fragments separated by electrophoresis

Restriction endonuclease enzyme cuts DNA strands into fragments of differing length at the restriction sites, one of which will contain the required gene.

restriction sites

required gene

restriction fragments

DNA fragments are separated by gel electrophoresis, using an agar gel (agarose gel):

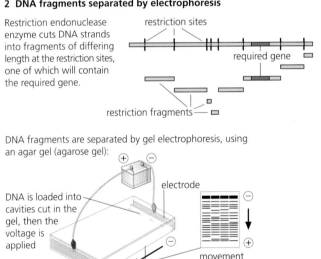

electrode

DNA is loaded into cavities cut in the gel, then the voltage is applied

electrode

agarose gel

movement of fragments

negative charge on DNA (due to phosphate groups) causes the fragments to move to the positive electrode (anode), but the gel has a 'sieving' effect – smaller fragments move more rapidly than larger ones

3 DNA transferred to nylon/nitrocellulose membrane (Southern blotting)

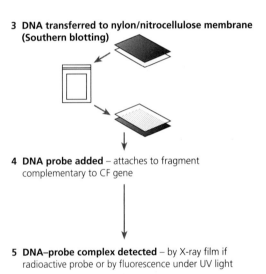

4 DNA probe added – attaches to fragment complementary to CF gene

5 DNA–probe complex detected – by X-ray film if radioactive probe or by fluorescence under UV light

Figure 2.54 Steps involved in genetic screening for cystic fibrosis.

Activity 2.17: Finding out more about cystic fibrosis

An approach to somatic gene therapy

Since the healthy CF gene codes for the protein that functions as an ion pump, the gene therapy initiative has involved getting copies of the healthy gene to the cells of the lung epithelia, delivered in an aerosol spray. The spray contains tiny lipid-bilayer droplets called liposomes to which copies of the healthy gene are attached. Liposomes fuse with plasma membranes and deliver the gene to epithelial cells. In trials, the treatment is effective, but only until the epithelial cells are routinely replaced. The treatment has to be regularly repeated. The cure can only be more permanent when it is targeted on the cells that make epithelial cells.

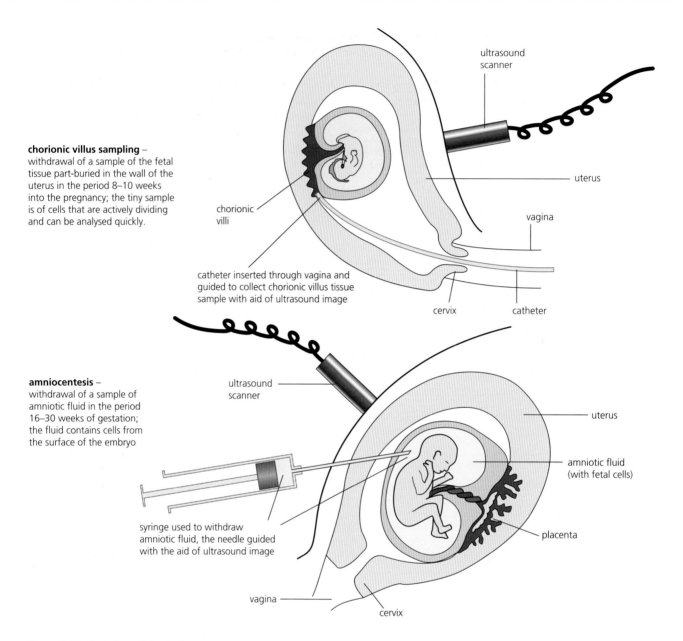

chorionic villus sampling – withdrawal of a sample of the fetal tissue part-buried in the wall of the uterus in the period 8–10 weeks into the pregnancy; the tiny sample is of cells that are actively dividing and can be analysed quickly.

chorionic villi

catheter inserted through vagina and guided to collect chorionic villus tissue sample with aid of ultrasound image

ultrasound scanner

uterus

vagina

cervix

catheter

amniocentesis – withdrawal of a sample of amniotic fluid in the period 16–30 weeks of gestation; the fluid contains cells from the surface of the embryo

ultrasound scanner

syringe used to withdraw amniotic fluid, the needle guided with the aid of ultrasound image

uterus

amniotic fluid (with fetal cells)

placenta

vagina

cervix

Figure 2.55 Screening a fetus in the uterus.

Genetic screening

The Human Genome Project (HGP), an initiative to map the entire human genome, was launched in 1990. The objectives were to discover the location of each human gene and the base sequence within its DNA structure. By June 2000, the sequencing of the human genome project had been achieved. At the same time as the HGP got underway, teams of scientists set about the sequencing of the DNA of other organisms, and more than 30 have been completed to date. Many aspects of biological investigation are being enhanced by the outcomes of this project.

Locating the cause of genetic disease

An outcome of the HGP is the ability to locate genes responsible for human genetic diseases. For example, the gene that codes for CFTR was sequenced in 1989. This has made possible the screening of human cells for the presence of the mutated cystic fibrosis gene (Figure 2.54). This may be applied to determine whether an individual is a carrier. It is possible for partners to be genetically screened to assess how likely they are to have children who will have cystic fibrosis, for example.

Another application involves the screening of embryos prepared for *in vitro* **fertilisation (IVF)**. In mammals, fertilisation – the fusion of male and female gametes to form a zygote – normally occurs in the upper part of the oviduct. However, not all partners can achieve this; either the male or female or both may be infertile, due to a number of possible causes. In some cases, a couple's infertility may be overcome by carrying out the fertilisation of eggs outside the body (*in vitro*). The key steps in IVF involve obtaining egg cells, isolating them from surrounding follicle cells, and then mixing them with healthy sperm cells. If fertilisation occurs, the fertilised egg cells are incubated so that embryos at the eight-cell stage may be placed in the uterus. If one (or more) becomes embedded there, then a normal pregnancy may follow. This procedure makes it possible for embryos to be screened for the presence of faulty genes prior to implantation.

Finally, screening for cystic fibrosis may be undertaken on cells obtained from a fetus in the uterus, either by **chorionic villus sampling** or by **amniocentesis** (Figure 2.55). Since it may be possible to abort a fetus that carries a genetic defect detected in this way, the practice generates ethical issues for parents, doctors and the whole medical team involved, and for society. These are considered below.

Social and ethical issues of genetic screening

Ethics are the moral principles that we feel ought to govern or influence the conduct of a society. They are concerned with how we decide what is right and what is wrong. Society consists of the totality of people living together in a country or region, forming a more-or-less ordered community. The term generally implies common purposes and interests in a majority of the members of that community. Communities and societies evolve traditions in their actions and views which come to be recognised as their **culture**. As societies become more complex, their culture may change and develop, and new ethical issues emerge. Today, developments in science and technology influence many aspects of people's lives. This certainly throws up increasingly complex ethical issues, which we need to examine and respond to. Genetic screening is just one case in point.

The process of making ethical decisions

First, we can agree what ethics is *not*. In Table 2.6 are listed aspects of our lives and culture that are all concerned with right and wrong in various ways. We may well be influenced by one or more, but they are not themselves ethics.

Table 2.6 Defining ethics by recognising what it is not.

Ethics is not ...	
... the law	Our laws are made by governments and may or may not be ethically based. (Sometimes, laws may be based solely on a view held by a powerful clique.)
... religion	In our society, there are followers of different (and conflicting?) religions and of none, yet ethics applies to us all. Religions may not address current issues quickly enough.
... cultural norms	Many of these are little more than 'fashions' that seemed acceptable at one time, and which may still be held uncritically.
... science	While science may seek to give us an understanding of the origins of our world and life, and how these work, it may not suggest how we should act.
... our feelings or conscience	These are likely to be the products of our early environment, general outlook and individual experiences, and our temperament.

Can you think of examples of the aspects in Table 2.6 that you feel are (or have been in the past) not ethical? This is perhaps most difficult in the case of 'our feelings'.

Table 2.7 Ethical standards to apply in decision making.

Criteria to be met in determining an ethical position on an issue	Further detail
Rights – we have a duty to ensure these	Human rights have been recognised and articulated.
	E.g. the **United Nations Universal Declaration of Human Rights**: 'All human beings are born free and equal in dignity and rights. They are endowed with reason and conscience and should act towards one another in a spirit of brotherhood.'
	E.g. **European Convention on Human Rights**, which includes security, liberty, political, due process, welfare/economic, and group rights.
	E.g. Countries with a **Bill of Rights** written into their Constitution – as in the USA: 'We hold these truths to be self-evident . . .'.
Justice – a principle to guide actions	The principle that all people should be treated equally, i.e. the idea of fairness in our actions to each individual, is considered essential.
Utilitarianism – a belief that the overall benefits should be greater than the 'costs'	The principle of the greatest good for the greatest number, and with actions that generate minimal harm to others' interests. In effect, this involves a cost–benefit analysis.
Common good – the core conditions that are essential to the welfare of all	Where the life of a working, interacting community is inherently good, then all actions should prosper and support the common interest.
Virtue – the placing of importance on a need to live a 'good life'	Ethical actions necessarily respect and embody values like truth, honesty, courage, compassion, generosity, tolerance, integrity, fairness, self-control and prudence.
Other criteria?	*Do you or your peers have any other criteria about which the whole group can be convinced?*

Next, using the ethical standards proposed in Table 2.7, we can address the ethical issues presented by genetic screening in particular, in the context of developments in reproduction biology.

Do we generally agree on the criteria to be met in making an ethical decision (Table 2.7), and what broadly constitutes each?

Agreement on these is an essential first step.

Then, what is needed is a reasoned explanation for an ethical approach that ensures a consensus by all involved. This can be achieved by discussion and argument.

Individuals must state their views simply, supporting them with facts for which there is evidence. While this phase goes on, peers need to listen carefully and prepare questions which help them understand and which probe the case being made. All distinctive views, opinions or approaches to the issue under discussion need this exposure.

Preparing such a personal statement of view can best be achieved by selecting two or more of the listed ethical standards which appear to especially apply to the issue, and show how each is relevant and guides you to your opinion. So, for example, you might identify all parties directly involved in the issue and reason the 'rights' of each that must be respected. Similarly, you might calculate the 'costs' and the 'benefits' for a good society of the position you advocate.

Finally, when the complexities of the issue are in the public domain, the group needs to work towards a common view. Remember, this cannot automatically be the majority view – established by means of a vote, for example. It may well be that a minority view emerges as the most ethical response. In fact, it has been observed that the process of civilisation has sometimes involved the process of adopting the ideas of minorities!

DL
www

2 End-of-topic test

(full End-of-topic tests are available on the DL Student website)

1 Osmosis is a special case of diffusion.

 a Describe *two* ways diffusion and active transport differ as mechanisms of membrane transit.

 b Define osmosis.

 c Explain how the free movements of many water molecules are hindered in the presence of dissolved substances such as ions or glucose.

 d Why is the plasma membrane described as a partially permeable membrane?

 e When red cells were immersed in distilled water, the plasma membrane of these cells burst, but plant cells in the same environment were unharmed. Explain this phenomenon.   (10)

2 For the following examples of movements across plasma membranes, identify the precise mechanism that is described in each case.

 a In the lining of the small intestine, tiny droplets of partially digested lipids are taken into cells of the epithelium layer.

 b ATP molecules (cell energy currency) are made in the interior of tiny cell organelles, but pass out into the rest of the cytoplasm at membrane protein sites that open as pores in the presence of ATP.

 c Water molecules cross the lipid bilayer at permanently open channels.

 d Special macrophage cells take in and dispose of small debris particles that become attached to their external surface.   (8)

3 Read this introduction to amino acids, protein structure and enzyme action, and then answer the following questions.

> To understand how enzymes work, we need to understand the structure of proteins. A protein is a long, ribbon-like molecule made by joining together many subunits – amino acids. Each amino acid has a central carbon atom, attached to which is an amino group ($-NH_2$), a carboxyl group ($-COOH$), and a side chain generally represented by R. The amino and acidic groups are involved in joining the amino acids to one another to form the backbone of the protein chain.
>
> The protein molecule does not lie flat like a ribbon – far from it. In watery environments, such as inside living cells, a protein chain will fold into a unique three-dimensional shape so that its hydrophobic regions are (generally) towards the outside. Weak bonds form between R groups and

these help stabilise the delicate shape. The shape of the enzyme molecule has to be just right if the enzyme is to function. If the enzyme molecule is deformed, even slightly, it will not work. Each different enzyme has its own highly specific shape with a pocket – the active site. It is here the substrate will fit.

> Only a few of the amino acids in the chain of an enzyme protein are involved in catalysis. These catalytic amino acids are not adjacent to one another in the chain. Instead they are quite spaced out along the length of the protein molecule. When the molecule folds up into its three-dimensional shape, however, the specific folding brings all the catalytic amino acid residues together in the active site.
>
> Once the substrate has fitted into the active site, the reaction is catalysed. Generally, the active site is lined with hydrophobic R groups and it contains groups that bind the substrate and other groups that catalyse the reaction. When the substrate binds, it disturbs the delicate balance of the flexible protein chain of the enzyme, causing it to alter its shape. This is why we refer to an induced fit of substrate to enzyme.

 a Show how two amino acids join together (line 9) by writing an equation for the condensation reaction forming a dipeptide, using a generalised structural formula for the two amino acids. Identify and label the bond formed.

 b In this passage, how does the author describe the primary structure of a protein?

 c What do you understand by a 'hydrophobic' region of a protein (line 16)?

 d Name or describe two types of weak bond that may form between different parts of the protein chain (line 17), stabilising the folded structure of the protein.

 e Enzymes are highly specific in their action; generally an enzyme reacts (combines) with only one type of substrate molecule. By reference to the description of enzyme structure given (lines 21–33), how can the specificity of enzyme action can be explained?

 f Once an 'induced fit' state (enzyme–substrate complex) has been achieved (line 42), an instantaneous reaction follows before the enzyme becomes ready to catalyse another reaction. Summarise the events of an enzyme-catalysed reaction in a simple equation, where E = enzyme, S = substrate, and Pr = product.   (12)

3 The voice of the genome

STARTING POINTS

■ Cells consist of cytoplasm surrounded by a plasma membrane and contain tiny sub-structures called organelles. The nucleus contains the chromosomes, and is the largest organelle in the cell. Observation of organelles requires the use of an electron microscope.

■ The cells of most multicellular organisms, including the mammals and flowering plants, are adapted to perform specialised functions.

■ Reproduction, the production of new individuals, is a characteristic of living things. In asexual reproduction, the offspring are genetically identical to the parent, but in sexual reproduction new individual(s), genetically different from the parents, are formed.

■ When cells divide, the nucleus divides first, and each new cell receives a nucleus. The nucleus controls and directs the activities of the cell throughout life, and contains the hereditary material which is passed to the next generation during reproduction.

■ In protein synthesis, the genetic code of part of a chromosome is transcribed into a single strand of messenger RNA (mRNA) which passes out into the cytoplasm. Here, the sequence of triplets of bases in mRNA is translated into the sequence of amino acids that are combined together to form proteins. Many proteins are enzymes that control the reactions of metabolism, growth and development.

3.1 Cell ultrastructure

The development of our knowledge of cell structure and function makes a fascinating story. It began in the seventeenth century, in Europe, but intensified over the past century or so with the development of the component disciplines that make up cell biology, today. Many biologists contributed to the early development of **cytology** (the study of cell structure) in the years that followed the first reported observations of cells using simple light microscopes. The realisation that cells are the fundamental unit of structure and function in living things evolved gradually from these observations.

The three basic ideas of the **cell theory** are that:

- cells are the building blocks of structure in living things
- cells are the smallest unit of life
- all cells are derived from other cells (pre-existing cells) by division.

Then, as a result of a steadily accelerating pace of developments in genetics, biochemistry and microbiology in particular, two further concepts have been added:

- cells contain a blueprint (that is, information) for their growth, development and behaviour
- within cells are the sites of all the chemical reactions of life (metabolism).

In Chapter 2, we explored the structure and functions of the plasma membrane – the hugely important barrier that surrounds the cytoplasm of cells (page 45). Here, we are concerned with the tiny structures found within the cytoplasm itself, known as cell **organelles**. Apart from the nucleus, few cell organelles are large enough to be observed, even in outline, using the light microscope. Today, our knowledge of the fine detail of cell structure, known as cell **ultrastructure**, is based on use of the electron microscope. We start by considering this instrument.

Activity 3.1: Early microscopy and the origin of 'cell theory'

A* Extension 3.1: Magnification and resolution

Transmission electron microscopy

In the electron microscope (EM), a beam of electrons is used to produce a magnified image in much the same way as the optical microscope uses light. The electron beam is generated by an electron gun, and is focused by electromagnets, rather than glass lenses. In transmission electron microscopy of biological material, the electron beam is passed through an extremely thin section of tissue. Membranes and other structures present in the cells are stained with heavy metal ions, making them electron-opaque – they stand out as dark areas in the image. We cannot see electrons, so the beam is focused onto a fluorescent screen for viewing, or onto a photographic plate for permanent recording (Figure 3.1).

Because the electron beam has a much shorter wavelength than light rays, the resolving power of this microscope is much greater than the best light microscopes. Used with biological materials, the limit of resolution in transmission electron microscopy is about 5 nm (page 44). For the light microscope, the resolving power is only about 0.2 μm. (This means two objects less than 0.2 μm apart will be seen as one object with this instrument.) So, it is difficult to exaggerate the importance of electron microscopy in the establishment of our knowledge of cell ultrastructure.

Note that the printed image (photograph) from transmission electron microscopy is called a transmission electron micrograph – abbreviated to TEM in the caption to an image obtained in this way (see Figure 3.4, for example).

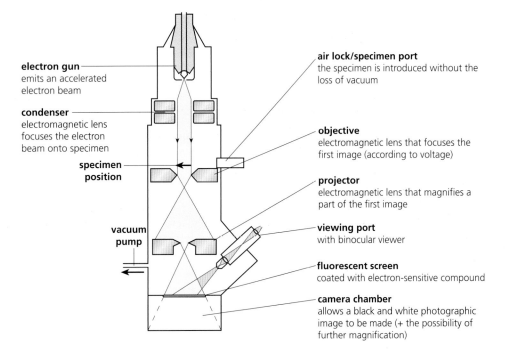

Figure 3.1 The transmission electron microscope.

electron gun
emits an accelerated electron beam

condenser
electromagnetic lens focuses the electron beam onto specimen

specimen position

vacuum pump

air lock/specimen port
the specimen is introduced without the loss of vacuum

objective
electromagnetic lens that focuses the first image (according to voltage)

projector
electromagnetic lens that magnifies a part of the first image

viewing port
with binocular viewer

fluorescent screen
coated with electron-sensitive compound

camera chamber
allows a black and white photographic image to be made (+ the possibility of further magnification)

Figure 3.2 An alternative form of electron microscopy produces scanning electron micrographs (SEM), which show surface detail.

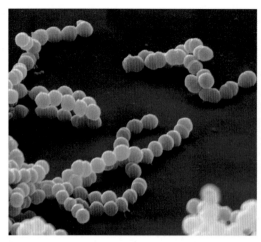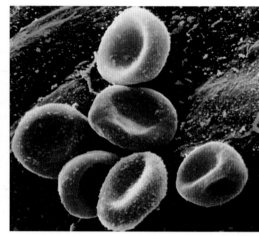

In scanning electron microscopy, the surface of the whole specimen is scanned by a beam of electrons. The three-dimensional image is created from electrons reflected from the surface and also from electrons generated there ('secondary electrons'). Larger specimens can be viewed by scanning electron microscopy than by transmission electron microscopy, but the resolution is not as great.

DL
A* Extension 3.2: The limitations of the electron microscope

Activity 3.2: Wider reading – 'Transmission electron microscopy'

HSW 3.1: Criterion 3 – The electron microscope as an appropriate methodology to answer scientific questions about the ultrastructure of cells

1 Explain the difference between *resolution* and *magnification*.

The EM and the discovery of two types of cell organisation

Living things were traditionally divided into two major groupings: animals and plants. However, the range of biological organisation has proved to be more diverse than this. For example, use of the electron microscope has disclosed two entirely different types of cellular organisation, based on the presence or absence of a nucleus.

Cells of organisms like the plants, animals and fungi have a large, obvious nucleus. The surrounding cytoplasm contains many different membranous organelles. These types of cells are called **eukaryotic** cells (meaning 'good nucleus'). We will examine this type of cell organisation first.

On the other hand, bacteria contain no true nucleus and their cytoplasm does not have the organelles of eukaryotes. These are called **prokaryotic** cells (meaning 'before the nucleus'). There are vast numbers of prokaryotes, and their evolutionary history extends further back: the ancestors of these organisms were the first forms of life to become established.

Another key difference between the cells of the prokaryotes and eukaryotes is their size. Prokaryotic cells are exceedingly small – about the size of individual organelles found in the cells of eukaryotes. We return to the examination of the structure of the prokaryotic cell later (page 105).

The ultrastructure of an animal cell

2 What is the difference between a *double membrane* and a *lipid bilayer*?

Today, the eukaryotic cell is seen as a 'bag' of organelles, many (but not all) of which are made of membranes. The fluid around the organelles is a watery (aqueous) solution of chemicals, called the **cytosol**. The chemicals in the cytosol are substances formed and used in the chemical reactions of life. All the reactions of life are known collectively as **metabolism**, and the chemicals are known as metabolites. The cytosol and organelles are contained within a special membrane, the **plasma membrane**. The ways in which this is crossed by all the metabolites that move between the cytosol and the environment of the cell have already been discussed (page 48). Now we shall consider the structure and function of the organelles of an animal cell.

Our picture of the arrangement of organelles within the plasma membrane of the cell has been built up by the examination of numerous transmission electron micrographs. This detailed picture, referred to as the ultrastructure of cells, is represented diagrammatically in Figure 3.3. A TEM of a particular animal cell is shown in Figure 3.4.

DL
Activity 3.3: Investigating the sizes of organelles

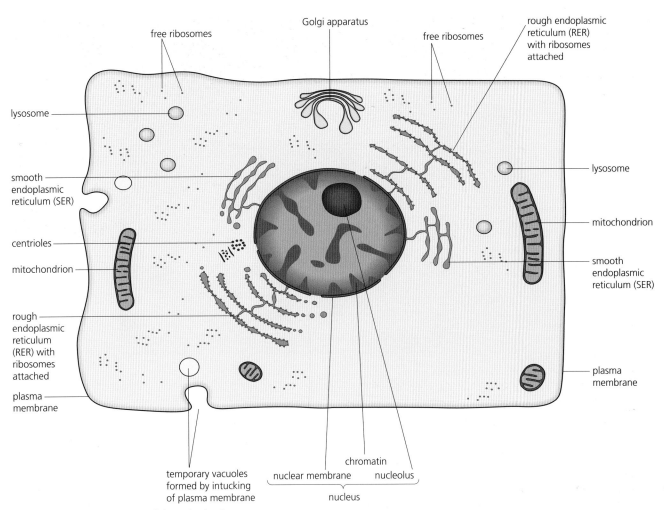

Figure 3.3 The ultrastructure of the animal cell.

Introducing the organelles

Nucleus

The **nucleus** is the largest organelle in the eukaryotic cell, typically 10–20 μm in diameter. It is surrounded by a double membrane, which contains many pores, each only about 100 nm in diameter. Tiny though they are, these pores are so numerous that they make up about one third of the nuclear membrane's surface area. This suggests that communication between nucleus and cytoplasm is important.

The nucleus contains the **chromosomes**. These thread-like structures are visible at the time the nucleus divides (page 110). At other times, the chromosomes appear dispersed as a diffuse network, called **chromatin**. One or more nucleoli are present in the nucleus – tiny, rounded, darkly-staining bodies. These are the site where ribosomes are synthesised. Chromatin, chromosomes and the nucleolus are visible only if stained with certain dyes. The everyday role of the nucleus in protein synthesis has already been discussed (page 77).

Mitochondria

Mitochondria appear mostly as rod-shaped or cylindrical organelles in electron micrographs, although occasionally their shape is more variable. They are relatively large organelles, typically 0.5–1.5 μm wide, and 3.0–10.0 μm long. Mitochondria are found in all cells, and are usually present in very large numbers – metabolically very active cells such as muscle fibres and hormone-secreting cells contain thousands of them in their cytoplasm.

Figure 3.4 TEM of a mammalian liver cell, with interpretive drawing.

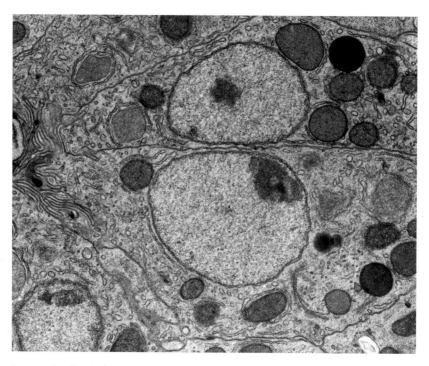

interpretive drawing

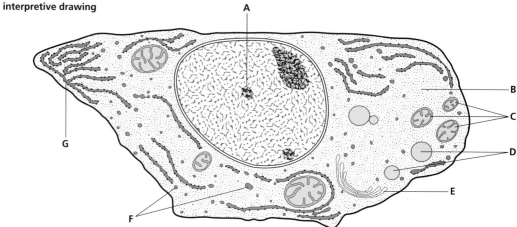

3 Identify the organelles labelled A to G in Figure 3.4, and state the chief role of each in the living cell.

The mitochondrion, like the nucleus, has a double membrane. Here, the outer membrane is a smooth boundary, while the inner membrane is infolded to form **cristae** (singular crista). The interior of the mitochondrion contains an aqueous solution of metabolites and enzymes, called the matrix. The mitochondrion is the site of the aerobic stages of respiration.

Figure 3.5 The mitochondrion.

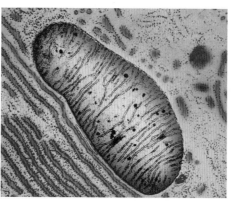

TEM of a thin section of a mitochondrion

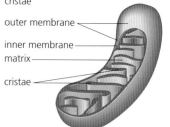

stereogram of a mitochondrion, cut open to show the inner membrane and cristae

outer membrane
inner membrane
matrix
cristae

In the mitochondrion, many of the enzymes of respiration are housed, and the 'energy currency' molecules (adenosine triphosphate, ATP) are formed.

Figure 3.6 The ribosome.

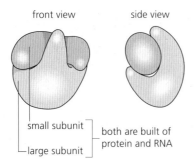

front view   side view

small subunit
large subunit
both are built of protein and RNA

Ribosomes

Ribosomes are the site of protein synthesis. They are minute structures, approximately 25 nm in diameter. They are built of two subunits, and do not have membranes as part of their structure. They are constructed of protein and the nucleic acid RNA. Many types of cell contain vast numbers of ribosomes.

Endoplasmic reticulum (ER)

Endoplasmic reticulum consists of a network of folded single membranes forming interconnected sheets, tubes or sacs. The origin of endoplasmic reticulum is the outer membrane of the nucleus, to which it may remain attached. The cytoplasm of metabolically active cells is commonly packed with endoplasmic reticulum, of which two distinct types are recognised.

- **Rough endoplasmic reticulum (RER)** has ribosomes attached to its outer surface. Vesicles are formed from swellings at the margins that become pinched off. A vesicle is a small, spherical organelle bounded by a single membrane, which is used to store and transport substances around the cell. For example, RER is the site of synthesis of proteins that are 'packaged' in vesicles, and are typically discharged from the cell, such as the digestive enzymes.
- **Smooth endoplasmic reticulum (SER)** has no ribosomes. SER is the site of synthesis of substances needed by cells. For example, SER is important in the manufacture of lipids. In the cytoplasm of voluntary muscle fibres, a special form of SER is the site of storage of calcium ions which have an important role in the contraction of muscle fibres.

TEM of RER

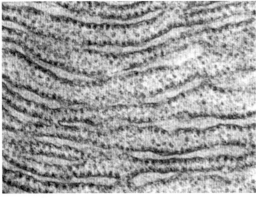

TEM of SER

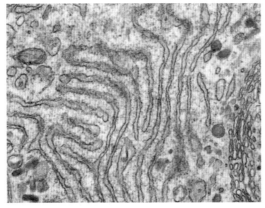

SER and RER in cytoplasm, showing origin from outer membrane of nucleus

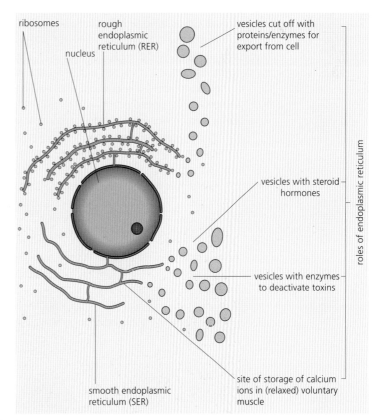

ribosomes
rough endoplasmic reticulum (RER)
nucleus
vesicles cut off with proteins/enzymes for export from cell
vesicles with steroid hormones
vesicles with enzymes to deactivate toxins
roles of endoplasmic reticulum
smooth endoplasmic reticulum (SER)
site of storage of calcium ions in (relaxed) voluntary muscle

Figure 3.7 Endoplasmic reticulum, rough (RER) and smooth (SER).

Golgi apparatus

The **Golgi apparatus** consists of a stack-like collection of flattened membranous sacs. One side of the stack of membranes is formed by the fusion of membranes of vesicles from the endoplasmic reticulum. At the opposite side of the stack, vesicles are formed from swellings at the margins that become pinched off.

The Golgi apparatus occurs in all cells, but it is especially prominent in metabolically active cells, such as secretory cells. It is the site of synthesis of specific biochemicals (like hormones, enzymes or polysaccharide macromolecules) which are then packaged into vesicles. In animal cells these vesicles may form lysosomes (see below).

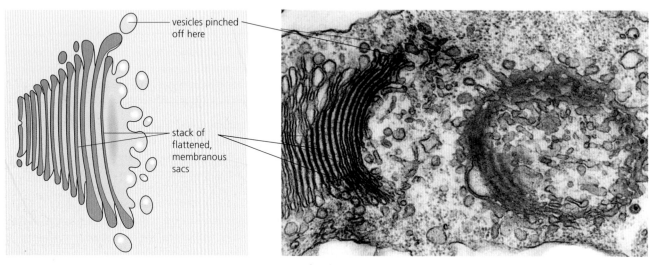

vesicles pinched off here

stack of flattened, membranous sacs

TEM of Golgi apparatus, in section and surface view

Figure 3.8 The Golgi apparatus.

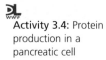

Activity 3.4: Protein production in a pancreatic cell

Lysosomes

Lysosomes are small spherical vesicles bound by a single membrane. They contain a concentrated mixture of hydrolytic ('digestive') enzymes, which are produced in the Golgi apparatus or by the RER.

Lysosomes are involved in the breakdown of the contents of imported food vacuoles – for example, a harmful bacterium that has invaded the body and been engulfed by one of the body's defence cells. It is then broken down, and the products of digestion escape into the liquid of the cytoplasm. Lysosomes also fuse with and digest any broken-down organelles in the cytoplasm. When an organism dies, the hydrolytic enzymes in the lysosomes of the cells escape into the cytoplasm and cause self-digestion (autolysis).

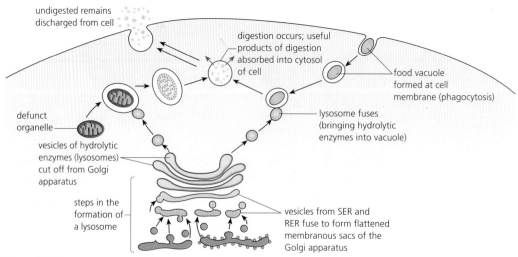

undigested remains discharged from cell

digestion occurs; useful products of digestion absorbed into cytosol of cell

food vacuole formed at cell membrane (phagocytosis)

defunct organelle

vesicles of hydrolytic enzymes (lysosomes) cut off from Golgi apparatus

lysosome fuses (bringing hydrolytic enzymes into vacuole)

steps in the formation of a lysosome

vesicles from SER and RER fuse to form flattened membranous sacs of the Golgi apparatus

Figure 3.9 Lysosomes.

4 What is the functional relationship between the ER, Golgi apparatus and lysosomes?

Microtubules and centrioles

Microtubules are straight, unbranched hollow cylinders, only 25 nm wide. They are commonly seen in the cytoplasm of eukaryotic cells. These tubes are made of a globular protein called tubulin, and are built up and broken down as needed by cells. Microtubules are involved in movements of cell components within the cytoplasm, acting to guide and direct organelles. The spindle fibres which appear during the nuclear division are microtubules, and they are responsible for the movement of chromosomes to the poles of the spindle.

Microtubules also form the **centrosome** which is found in animal cells besides the nucleus. The centrosome comprises two **centrioles** lying at right angles to each other; each consists of nine triplets of microtubules (Figure 3.10). Centrioles separate and move to opposite ends of the nucleus before nuclear division (page 110).

Microtubules are also found in the motile organelles known as cilia and flagella that occur on some cells. They also occur in the midpiece of a sperm cell (page 120).

Figure 3.10 The centrosome, of two centrioles lying at right angles, occurs just outside the nucleus (double membrane).

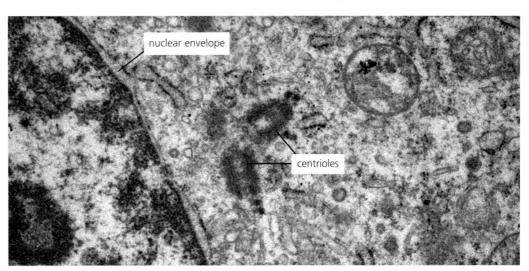

nuclear envelope

centrioles

5 How has the electron microscope increased our knowledge of cell structure?

■ **Extension:** Non-cellular organisation – an exceptional condition

In addition to the familiar unicellular and multicellular organisation of living things, there are a few examples of **multinucleate** organs (and organisms) without divisions into separate cells. This type of arrangement is known as **acellular** organisation.

An example of an acellular organ is the striped muscle fibre 'cells' that make up the skeletal muscles of mammals. Each elongated fibre contains several nuclei. (This tissue is examined in A2 Biology.)

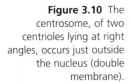

Activity 3.5: Introducing the British Society for Cell Biology, and Centre of the Cell websites

Prokaryotic cell structure

The **prokaryotes** are the bacteria and cyanobacteria (photosynthetic bacteria). These micro-organisms, typically unicellular, have a fundamentally different cell structure from eukaryotes. The organelles of eukaryotic cells, introduced above, have no equivalents in the cells of prokaryotes like the bacteria. Here the most elaborate structures in the cytoplasm are ribosomes, functionally similar to those of eukaryotes, but smaller. Also present in some bacteria are modest 'in-tuckings' of the cell membrane, forming simple membrane systems on which particular enzymes or pigments may be attached.

Typically, prokaryotic cells are about the size of organelles like mitochondria and chloroplasts in eukaryotic cells. All these features are illustrated in Figure 3.11.

Figure 3.11 The structure of a bacterium.

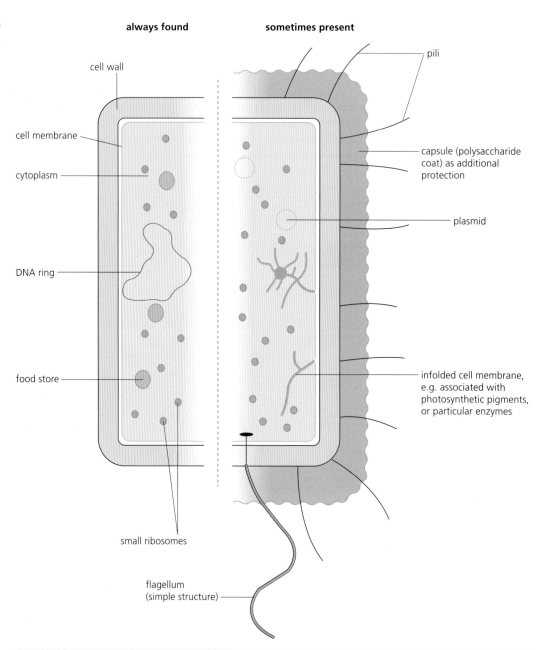

always found   sometimes present

cell wall

cell membrane

cytoplasm

DNA ring

food store

pili

capsule (polysaccharide coat) as additional protection

plasmid

infolded cell membrane, e.g. associated with photosynthetic pigments, or particular enzymes

small ribosomes

flagellum (simple structure)

Activity 3.6: Prokaryotic and eukaryotic cells compared

A* Extension 3.3: The evidence for the cell theory

■ Extension: A possible origin for mitochondria and chloroplasts

Present-day prokaryotes are similar to fossil prokaryotes, some of which are 3500 million years old. By comparison, the earliest eukaryotic cells date back only 1000 million years. Thus eukaryotes must have evolved, surrounded by prokaryotes that were long-established organisms. It is possible that, in the evolution of the eukaryotic cell, prokaryotic cells – which at one stage were taken up into food vacuoles for digestion, for example – came to survive as organelles inside the host cell, rather than becoming food items! They might then have become integrated into the biochemistry of their 'host' cell, with time.

If this hypothesis is correct, it would explain why mitochondria (and chloroplasts) contain a ring of DNA double helix, just as a bacterial cell does. They also contain small ribosomes like those of prokaryotes. These features have caused some evolutionary biologists to suggest that these organelles may be descendants of free-living prokaryotic organisms that came to inhabit larger cells. It may seem a fanciful idea, but not impossible.

3.2 Cells make organisms

Amoeba (page 57), *Chlorella* and a bacterium (Figure 3.11) are all examples of unicellular organisation. Unicellular organisms are structurally simple, but they are able to perform all the functions and activities of life within a single cell. The cell feeds, respires, excretes, is sensitive to internal and external conditions (and may respond to them), may move, and eventually divides or reproduces.

All other organisms are multicellular. Multicellular organisms are a diverse group – a few are little more than a colony of identical cells that appear to have remained together after

Two of the many tissues that make up a leaf are illustrated here.

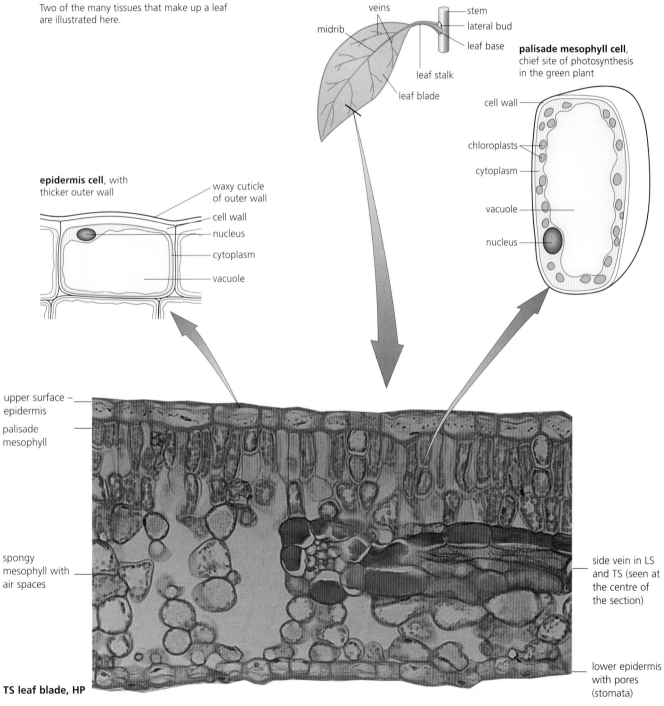

TS leaf blade, HP

Figure 3.12 Tissues of a leaf.

division. An example of this is a colony of yeast (*Saccharomyces*) cells that have failed to separate.

However, the majority of multicellular organisms are like the mammals (and flowering plants) in that they grow and develop from a zygote, formed by fusion of male and female gametes, into a fully grown or adult structure. Most animals have a limited period of growth and development, resulting in an adult organism of characteristic size and form. The adult organism has cells that are mostly highly specialised to perform particular functions. Specialised cells are efficient in the conduct of a particular function, such as transport, or support, or protection. At the same time, specialised cells do not have the facility to do other things. We say the resulting differences between cells are due to **division of labour**.

There is an obvious trade-off between the advantages of specialisation and the condition of dependency. It seems that in this type of specialisation, increased efficiency is achieved at a price. Here the specialised cells are now totally dependent on the activities of other cells. For example, in animals, nerve cells are adapted for the transport of nerve impulses, but are dependent on blood cells for oxygen, and on heart muscle cells to pump the blood.

Another important feature of specialised cells in multicellular organisms is that they become organised into tissues and organs during development.

6 Give one example each of an organ, a tissue and an undifferentiated cell to be found in:
a a flowering plant
b a mammal.

- A **tissue** is a group of similar cells specialised to perform a particular function (for example, muscle tissue of mammals, mesophyll tissue of flowering plant leaves, Figure 3.12).
- An **organ** is a collection of different tissues which performs a specialised function (for example, a leaf of a plant, the gut of a mammal, Figure 3.13).

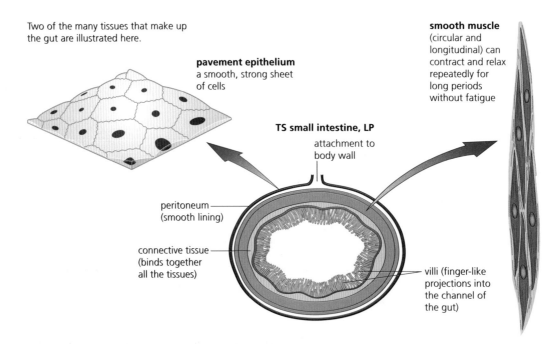

Two of the many tissues that make up the gut are illustrated here.

pavement epithelium
a smooth, strong sheet of cells

smooth muscle (circular and longitudinal) can contract and relax repeatedly for long periods without fatigue

TS small intestine, LP

attachment to body wall

peritoneum (smooth lining)

connective tissue (binds together all the tissues)

villi (finger-like projections into the channel of the gut)

Figure 3.13 Tissues of part of the mammalian gut.

Activity 3.7: Wider reading – 'Developmental biology'

Incidentally, the study of tissues and the ways they combine together to make organisms is a branch of biology called **histology**. However, before we can consider this aspect of development we need to discuss the cell division processes that make development possible.

Cell division and the cell cycle

To recap, multicellular organisms begin life as a single cell which grows and divides. During growth, this cycle is repeated seemingly endlessly, forming many cells. It is these cells that eventually make up the adult organism. So, new cells arise by division of existing cells, and the cycle of growth and division is called the **cell division cycle**. This cycle has three main stages:

■ interphase
■ division of the nucleus by a process (mitosis) that results in two nuclei, each with identical sets of chromosomes
■ division of the cytoplasm and whole cell (known as cytokinesis).

In fact, in each stage of the cell cycle particular events occur. These events are summarised in Figure 3.14, and they are discussed below.

Look at the sub-division of interphase – distinctive features are identified in each.

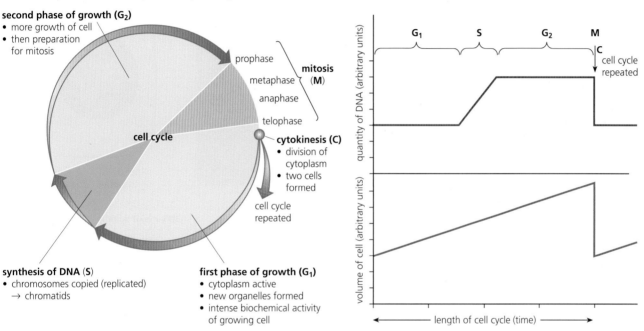

the cell cycle consists of interphase and mitosis

interphase = $G_1 + S + G_2$

change in cell volume and quantity of DNA during a cell cycle

second phase of growth (G_2)
• more growth of cell
• then preparation for mitosis

prophase
metaphase
anaphase
telophase
mitosis (M)

cell cycle

cytokinesis (C)
• division of cytoplasm
• two cells formed

cell cycle repeated

synthesis of DNA (S)
• chromosomes copied (replicated) → chromatids

first phase of growth (G_1)
• cytoplasm active
• new organelles formed
• intense biochemical activity of growing cell

quantity of DNA (arbitrary units)

G_1 S G_2 M
C
cell cycle repeated

volume of cell (arbitrary units)

length of cell cycle (time)

Figure 3.14 The stages of the cell cycle.

Interphase

Interphase is always the longest part of the cell cycle, but it is of extremely variable length. When growth is fast, as in a developing human embryo, or the growing point of a young green plant stem, interphase may last about 24 hours or less. On the other hand, in mature cells that infrequently divide it lasts a very long period – possibly indefinitely. For example, some cells, once they have differentiated, rarely or never divide again. Here the nucleus remains at interphase permanently.

An overview of interphase

When the living nucleus of a cell at interphase is observed by light microscopy, for example, the nucleus appears to be 'resting'. This is not the case. During interphase, the chromosomes are actively involved in protein synthesis. From the chromosomes, copies of the information of particular genes or groups of genes (in the form of mRNA – page 77) are taken for use in the cytoplasm. It is in the ribosomes of the cytoplasm that proteins are assembled from amino acids, combining them in sequences dictated by the information from the gene, in the form of mRNA.

The distinctive chromosomes, visible during mitosis (Figure 3.17), become dispersed in interphase. They are now referred to as chromatin. Amongst the chromatin can be seen one or more dark-staining structures, known as nucleoli (singular nucleolus). Chemically, the nucleoli consist of protein and RNA, and they are the site of synthesis of the ribosomes. These tiny organelles then migrate out into the cytoplasm.

For simplicity, the drawings show mitosis in a cell with a single pair of homologous chromosomes.

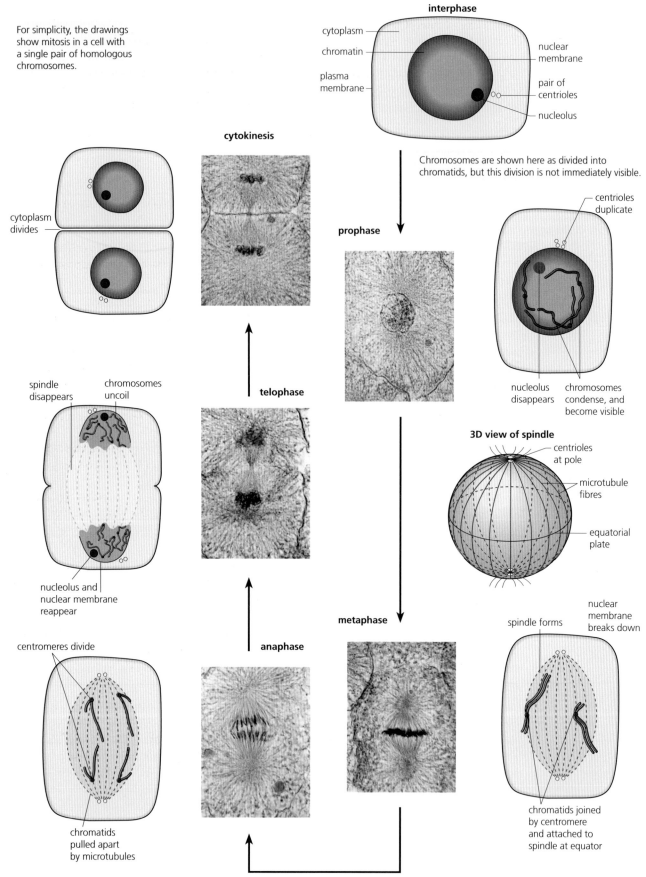

interphase

cytoplasm

chromatin

plasma membrane

nuclear membrane

pair of centrioles

nucleolus

Chromosomes are shown here as divided into chromatids, but this division is not immediately visible.

cytokinesis

cytoplasm divides

prophase

centrioles duplicate

nucleolus disappears

chromosomes condense, and become visible

telophase

spindle disappears

chromosomes uncoil

nucleolus and nuclear membrane reappear

3D view of spindle

centrioles at pole

microtubule fibres

equatorial plate

centromeres divide

anaphase

metaphase

spindle forms

nuclear membrane breaks down

chromatids pulled apart by microtubules

chromatids joined by centromere and attached to spindle at equator

Figure 3.15 Mitosis in an animal cell.

The stages of interphase

During the first phase of growth (G_1), the synthesis of new organelles takes place in the cytoplasm. There is also a time of intense biochemical activity in the cytoplasm and organelles, and there is an accumulation of stored energy before nuclear division occurs again.

Next is a period of synthesis of DNA (S), when each chromosome makes a copy of itself. It is said to replicate. The two identical structures formed are called **chromatids**. The chromatids remain attached until they divide during mitosis.

Finally, there is a second phase of growth (G_2), which is a continuation of the earlier time of intense biochemical activity and increase in the amount of cytoplasm.

Mitosis

When cell division occurs, the nucleus divides first. In mitosis, the chromosomes, present as the chromatids formed during interphase, are separated and accurately and precisely distributed to two daughter nuclei.

Here, mitosis is presented and explained as a process in four phases (Figure 3.15), but this is for convenience of description only. Mitosis is one continuous process with no breaks between the phases.

- In **prophase** the chromosomes become visible as long thin threads. They increasingly shorten and thicken by a process of supercoiling (Figure 3.16). Only at the end of prophase is it possible to see that they consist of two chromatids held together at the **centromere**. At the same time, the nucleolus gradually disappears and the nuclear membrane breaks down.
- In **metaphase** the centrioles move to opposite ends of the cell (in an animal cell). Microtubules of the cytoplasm start to form into a spindle, radiating out from the centrioles. Microtubules attach to the centromeres of each pair of chromatids, and these are arranged at the equator of the spindle. (Note that in plant cells, a spindle of exactly the same structure is formed, but without the presence of the centrioles.)
- In **anaphase** the centromeres divide, the spindle fibres shorten, and the chromatids are pulled by their centromeres to opposite poles. Once separated, the chromatids are referred to as chromosomes.
- In **telophase** a nuclear membrane re-forms around both groups of chromosomes at opposite ends of the cell. The chromosomes 'decondense' by uncoiling, becoming chromatin again. The nucleolus re-forms in each nucleus. Interphase follows division of the cytoplasm.

Cytokinesis

Division of the cytoplasm, known as **cytokinesis**, follows telophase. During division, cell organelles such as mitochondria and chloroplasts become distributed evenly between the cells. In animal cells, division is by in-tucking of the plasma membrane at the equator of the spindle, 'pinching' the cytoplasm in half (Figure 3.15).

In plant cells, the Golgi apparatus forms vesicles of new cell wall materials which collect along the line of the equator of the spindle, known as the cell plate. Here the vesicles coalesce forming the new plasma membranes and cell walls between the two cells (Figure 3.17).

Observing chromosomes during mitosis

Actively dividing cells, such as those at the growing points of the root tips of plants, include many cells undergoing mitosis. This tissue can be isolated, stained with an orcein ethanoic (aceto-orcein) stain, squashed, and then examined under the high-power objective of the microscope. With this stain, nuclei at interphase appear red–purple with almost colourless cytoplasm, but the chromosomes in cells undergoing mitosis will be more darkly stained, rather as they appear in the photomicrographs in Figure 3.15. The procedure is summarised in the flow diagram in Figure 3.18.

---

7 What structures of the interphase nucleus can be seen by electron microscopy?

8 Suggest a main advantage of chromosomes being 'supercoiled' during metaphase of mitosis.

DL
www
A* Extension 3.4: The different roles of proteins in the nucleus

■ Extension: The packaging of DNA in the chromosomes

The total length of the DNA of human chromosomes is over 2 m, shared out between the 46 chromosomes. Each chromosome contains one, very long DNA molecule. Of course, chromosomes are of different lengths, but we can estimate that, for a typical chromosome of 5 μm length, an approximately 5 cm long DNA molecule is contained within it (that is, about 50 000 μm of DNA packed into 5 μm of chromosome!). Today we know that, while some of the proteins of the chromosome are enzymes involved in the copying and repair reactions of DNA, the bulk of chromosome protein has a support and packaging role for DNA.

One sort of packaging protein is a substance called histone. This is a basic (positively charged) protein containing a high concentration of amino acid molecules with additional base groups ($-NH_2$), such as lysine and arginine. These histones occur clumped together, and provide support to the lengths of the DNA double helix that occur wrapped around them, giving the appearance of beads on a thread. The 'bead thread' is itself coiled up, forming the chromatin fibre. The chromatin fibre is again coiled, and the coils are looped around a 'scaffold' protein fibre, made of a non-histone protein. This whole structure is folded ('supercoiled') into the much-condensed metaphase chromosome (Figure 3.16).

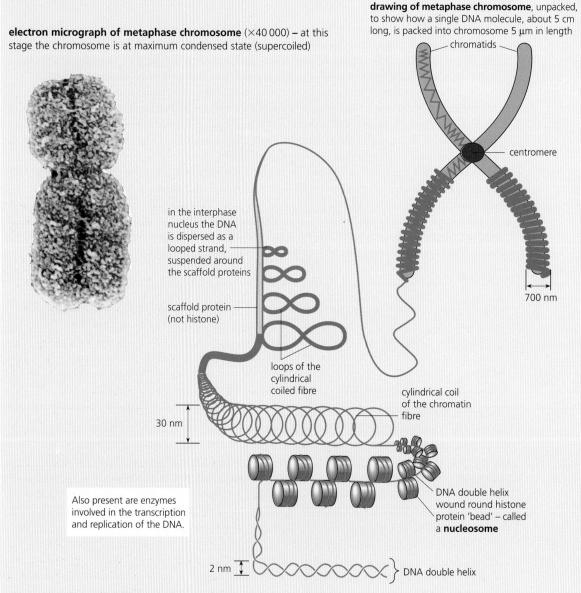

drawing of metaphase chromosome, unpacked, to show how a single DNA molecule, about 5 cm long, is packed into chromosome 5 μm in length

— chromatids

electron micrograph of metaphase chromosome (×40 000) – at this stage the chromosome is at maximum condensed state (supercoiled)

— centromere

700 nm

in the interphase nucleus the DNA is dispersed as a looped strand, suspended around the scaffold proteins

scaffold protein (not histone)

loops of the cylindrical coiled fibre

cylindrical coil of the chromatin fibre

30 nm

Also present are enzymes involved in the transcription and replication of the DNA.

DNA double helix wound round histone protein 'bead' – called a **nucleosome**

2 nm

DNA double helix

Figure 3.16 The packaging of DNA in the chromosomes.

Figure 3.17 Cytokinesis in a plant cell.

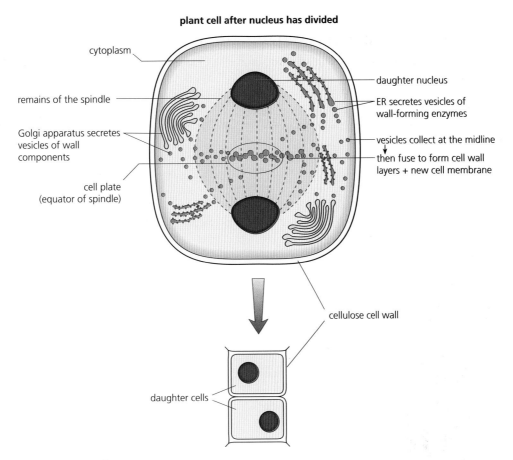

plant cell after nucleus has divided

cytoplasm

remains of the spindle

Golgi apparatus secretes vesicles of wall components

cell plate (equator of spindle)

daughter nucleus

ER secretes vesicles of wall-forming enzymes

vesicles collect at the midline
then fuse to form cell wall layers + new cell membrane

cellulose cell wall

daughter cells

The significance of mitosis

The significance of mitosis arises because the 'daughter' cells produced by this nuclear division have a set of chromosomes identical to each other and to the parent cell from which they were formed.

This occurs because:

- an exact copy of each chromosome is made by accurate replication (page 73) during interphase, when two chromatids are formed
- chromatids remain attached by their centromeres during metaphase of mitosis, when each becomes attached to a spindle fibre at the equator of the spindle
- centromeres then divide during anaphase and the chromatids of each pair are pulled apart to opposite poles of the spindle – thus, one copy of each chromosome moves to each pole of the spindle
- the chromosomes at the poles form the new nuclei – two to a cell at this point
- two cells are then formed by division of the cytoplasm at the midpoint of the cell, each with an exact copy of the original nucleus.

Where mitosis is commonly observed

In the growth and development of an embryo, it is essential that all cells carry the same genetic information as the existing cells from which they are formed, and the surrounding cells or tissues. Similarly, when replacement of damaged or worn out cells occurs, exact copies of the original cells are required. This is essential because otherwise different parts of our body might start working to conflicting blueprints. The results would be chaos!

Further, mitotic cell division is also the basis of all forms of **asexual reproduction**, where this occurs. Here, the offspring produced are identical to the parent. In this, asexual reproduction is completely different from sexual reproduction, where significant differences always arise between the individual offspring produced, and between offspring and their parents. We discuss sexual reproduction next.

Activity 3.8: Estimating the duration of the phases of mitosis

A* Extension 3.5: Programmed cell death

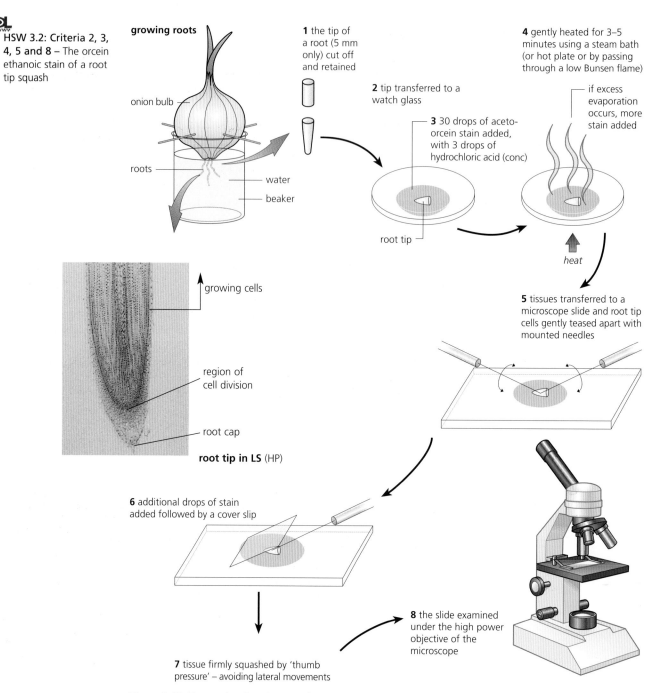

HSW 3.2: Criteria 2, 3, 4, 5 and 8 – The orcein ethanoic stain of a root tip squash

growing roots

onion bulb

roots

water

beaker

1 the tip of a root (5 mm only) cut off and retained

2 tip transferred to a watch glass

3 30 drops of aceto-orcein stain added, with 3 drops of hydrochloric acid (conc)

root tip

4 gently heated for 3–5 minutes using a steam bath (or hot plate or by passing through a low Bunsen flame)

if excess evaporation occurs, more stain added

heat

5 tissues transferred to a microscope slide and root tip cells gently teased apart with mounted needles

growing cells

region of cell division

root cap

root tip in LS (HP)

6 additional drops of stain added followed by a cover slip

7 tissue firmly squashed by 'thumb pressure' – avoiding lateral movements

8 the slide examined under the high power objective of the microscope

Figure 3.18 The orcein ethanoic stain of an onion root tip squash.

3.3 Organisms and reproduction

Reproduction is the production of new individuals by an existing member or members of the same species. It is a fundamental characteristic of living things; the ability to self-replicate in this way sets the living world apart from the non-living.

In reproduction, a parent generation effectively passes on a copy of itself in the form of the genetic material, to another generation, the offspring. The genetic material of an organism consists of its chromosomes, made of nucleic acid (page 70). Organisms reproduce either asexually or sexually and many reproduce by both these methods.

Asexual reproduction

In asexual reproduction, a single organism produces new individuals. Asexual means 'non-sexual'; no gametes are formed in asexual reproduction. The cells of the new offspring are produced by **mitosis**, so the progeny are identical to the parent and to each other. Incidentally, identical offspring are known as **clones**.

Organisms that can reproduce by asexual means often do so as soon as they become established in a new habitat or reach a certain size, provided they are well supplied with nutrients. An advantage of asexual reproduction is that a large number of new individuals are produced by a single parent, quickly. The outcome may be that a suitable habitat is successfully colonised and ideal growing conditions exploited quickly. The new individuals all have the qualities of the successful parent, since they are genetically identical.

Sexual reproduction

In sexual reproduction, two **gametes** (specialised sex cells) fuse to form a **zygote**, which then grows into a new individual. Fusion of gametes is called **fertilisation**. In the process of gamete formation a nuclear division known as **meiosis** halves the normal chromosome number (Figure 3.19). That is, gametes are **haploid**, and fertilisation restores the **diploid** number of chromosomes (Figure 3.20). Without the reductive nuclear division in the process of sexual reproduction, the chromosome number would double in each generation. Remember, the offspring produced by sexual reproduction show variations, in complete contrast with offspring formed by asexual reproduction. Mammals reproduce by sexual reproduction only, but many flowering plants reproduce by both asexual and sexual reproduction.

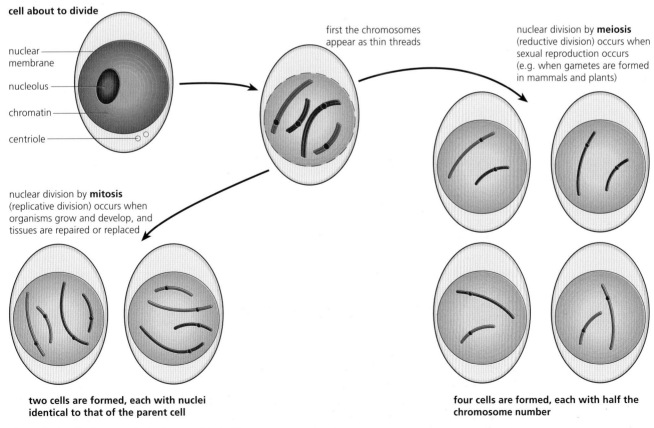

cell about to divide

nuclear membrane

nucleolus

chromatin

centriole

first the chromosomes appear as thin threads

nuclear division by **meiosis** (reductive division) occurs when sexual reproduction occurs (e.g. when gametes are formed in mammals and plants)

nuclear division by **mitosis** (replicative division) occurs when organisms grow and develop, and tissues are repaired or replaced

two cells are formed, each with nuclei identical to that of the parent cell

four cells are formed, each with half the chromosome number

Figure 3.19 Mitosis and meiosis, the significant differences.

Figure 3.20 Meiosis and the diploid life cycle.

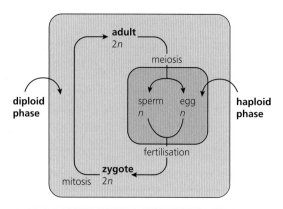

Self-fertilisation = sperm + egg from same individual.
Cross-fertilisation = sperm + egg from different individuals.

Meiosis, the reduction division

Meiosis is part of the life cycle of every organism that reproduces sexually. In meiosis, four daughter cells are produced, each with half the number of chromosomes of the parent cell. Halving of the chromosome number in gametes is essential, since at fertilisation the number is doubled.

How does meiosis work?

Meiosis involves two divisions of the nucleus, known as meiosis I and meiosis II, both of which *superficially* resemble mitosis. As in mitosis, chromosomes replicate to form chromatids during interphase, before meiosis occurs.

Then, early in meiosis I, homologous chromosomes pair up. By the end of meiosis I, homologous chromosomes have separated again, but the chromatids they consist of do not separate until meiosis II. Thus, **meiosis consists of two nuclear divisions but only one replication of the chromosomes.**

The process of meiosis

The key events in meiosis are summarised in Figure 3.21. In the interphase that precedes meiosis, the chromosomes are replicated as chromatids, but between meiosis I and II there is no further interphase, so no replication of the chromosomes occurs *during* meiosis.

As meiosis begins, the chromosomes become visible. At the same time, homologous chromosomes pair up. (Remember, in a diploid cell each chromosome has a partner that is the same length and shape and with the same linear sequence of genes. It is these partner chromosomes that pair up.)

When the homologous chromosomes have paired up closely, each pair is called a **bivalent**. Members of the bivalent continue to coil and shorten.

During the coiling and shortening process within the bivalent, the **chromatids** frequently break. Broken ends rejoin more or less immediately. When non-sister chromatids from homologous chromosomes break and rejoin they do so at exactly corresponding sites, so that a cross-shaped structure called a **chiasma** is formed at one or more places along a bivalent. The event is known as a **crossing over** because lengths of genes have been exchanged between chromatids.

Then, when members of the bivalents start to repel each other and separate, the bivalents are (initially) held together by one or more chiasmata. This temporarily gives an unusual shape to the bivalent. So, crossing over is an important mechanical event (as well as a genetic event).

Next the spindle forms. Members of the bivalents become attached by their centromeres to the fibres of the spindle at the equatorial plate of the cell. Spindle fibres pull the homologous chromosomes apart, to opposite poles, but the individual chromatids remain attached together by their centromeres.

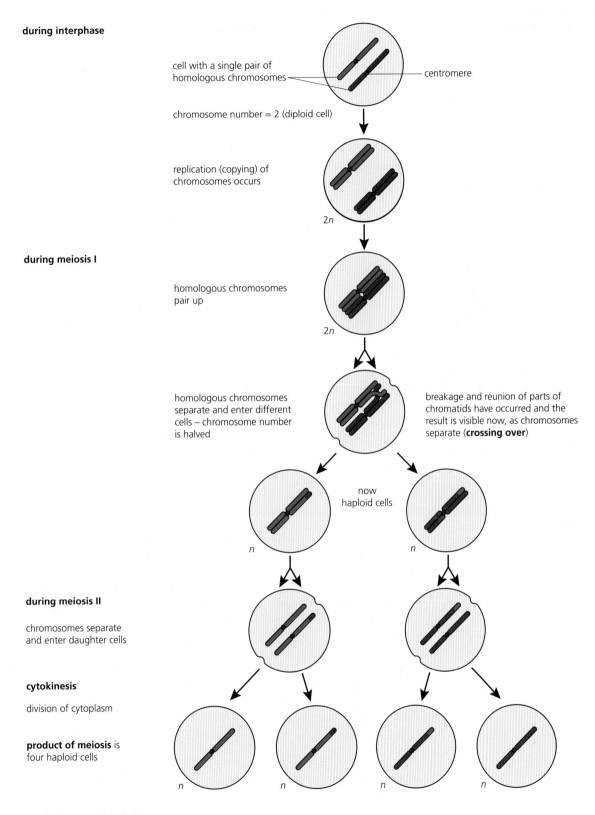

during interphase

cell with a single pair of homologous chromosomes

centromere

chromosome number = 2 (diploid cell)

replication (copying) of chromosomes occurs

2n

during meiosis I

homologous chromosomes pair up

2n

homologous chromosomes separate and enter different cells – chromosome number is halved

breakage and reunion of parts of chromatids have occurred and the result is visible now, as chromosomes separate (**crossing over**)

now haploid cells

n

n

during meiosis II

chromosomes separate and enter daughter cells

cytokinesis

division of cytoplasm

product of meiosis is four haploid cells

n

n

n

n

Figure 3.21 The process of meiosis.

9 Explain what is the major consequence of there being no interphase between meiosis I and meiosis II.

DL
www
Activity 3.9: Wider reading – 'Mechanisms of meiosis'

Meiosis I ends with two cells each containing a single set of chromosomes each made of two chromatids. These cells do not go into interphase, but rather continue smoothly into meiosis II. This takes place at right angles to meiosis I, but is exactly like mitosis. Centromeres of the chromosomes divide and individual chromatids now move to opposite poles. Now there are four cells, each with half the chromosome number of the original parent cell (haploid).

Meiosis and genetic variation

Meiosis is a major source of genetic variation – the haploid cells produced by meiosis differ from each other for two reasons.

- There is **independent assortment** of maternal and paternal homologous chromosomes. This happens because the way the bivalents line up at the equator of the spindle in meiosis I is entirely random. Which chromosome of a given pair goes to which pole is unaffected by (independent of) the behaviour of the chromosomes in other pairs. Orientation at the equator of the spindle is random. Independent assortment is illustrated in Figure 3.22, in a parent cell with a diploid number of 4 chromosomes. In human cells the number of pairs of chromosomes is 23. Here, the number of possible combinations of chromosomes that can be formed by

Figure 3.22 Genetic variation due to independent assortment.

Independent assortment is illustrated in a parent cell with two pairs of homologous chromosomes (four bivalents). The more bivalents there are, the more variation is possible. In humans, for example, there are 23 pairs of chromosomes giving over 8 million combinations.

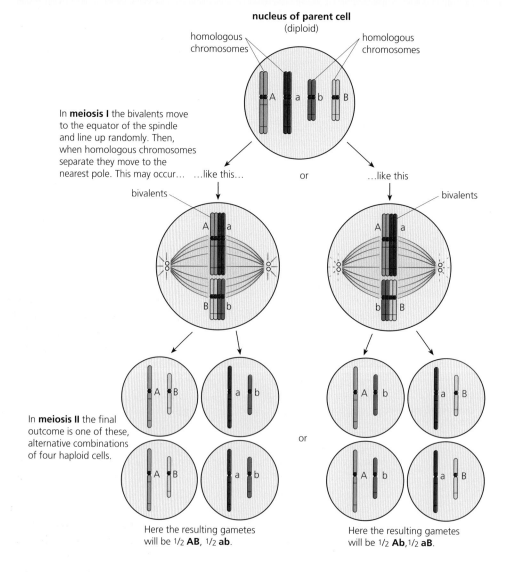

Activity 3.10: Mitosis and meiosis compared

random orientation during meiosis is 2^{23}, which is over 8 million. We see that independent assortment *alone* generates a huge amount of potential variation in the coded information carried by different gametes into the fertilisation stage.

- There is **crossing over** of segments of individual maternal and paternal homologous chromosomes. These events result in new combinations of genes on the chromosomes of the haploid cells produced (Figure 3.23).

Figure 3.23 Genetic variation due to crossing over between non-sister chromatids.

The effects of genetic variation are shown in one pair of homologous chromosomes.
Typically, two, three or more chiasmata form between the chromatids of each bivalent at prophase I.

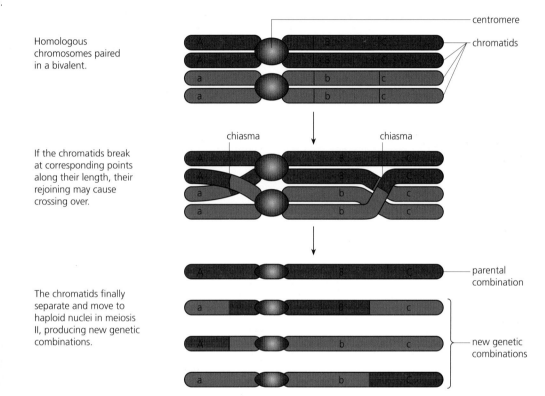

Homologous chromosomes paired in a bivalent.

If the chromatids break at corresponding points along their length, their rejoining may cause crossing over.

The chromatids finally separate and move to haploid nuclei in meiosis II, producing new genetic combinations.

10 What are the essential differences between mitosis and meiosis?

Fertilisation in mammals

Gametes – structure and function
Male and female gametes are distinctly different in structure and function.

The **sperms** are formed in the testes, in tiny sac-like structures, called seminiferous tubules, from puberty onwards. These tubules make up the bulk of the testes (Figure 2.52, page 91). Sperms are immobile when first formed, and are stored in the epididymis, prior to activation for ejaculation from the sperm ducts and urethra.

Look at the sperm cell shown in Figure 3.24.

A sperm is a cell adapted for reaching the egg cell (while at the secondary oocyte stage), partly as a result of its own motility. On reaching its target, a sperm penetrates between the surrounding follicle cells, and delivers the male nucleus within the egg cell. The tail, an organelle called a flagellum, propels the sperm in the phase of its journey when it moves by swimming. There are numerous mitochondria in the midpiece that provide ATP for this locomotion. The head consists of the haploid nucleus and the acrosome, a vesicle containing hydrolytic enzymes required for penetration.

The **egg cells** exist as primary oocytes in the ovaries at the time of birth of the female mammal, long before sexual maturation commences. At puberty, secondary oocytes form in the ovaries, and a few mature each month. Then, in the human female, one secondary oocyte is

A* Extension 3.6: Eukaryotic flagella and the structure of the sperm 'tail'

11 Calculate the magnifications of the sperm and of the secondary oocyte in the drawings in Figure 3.24.

released, is drawn into the oviduct funnel and begins to pass down the oviduct where fertilisation may occur.

A secondary oocyte contains a large egg cell (note the size comparison given in Figure 3.24), and is adapted for the process of becoming embedded (implanted) in the wall of the uterus, following fertilisation. Perhaps we ought to think of egg cells as originally containing a large stored food reserve at some time, prior to the evolution of internal fertilisation and internal development, where they are in contact with a supply of available food from the female mammal, via the placenta.

Figure 3.24 Human gametes at fertilisation.

sperm (♂)
2.5–3.5 μm wide, 5–7 μm long

secondary oocyte (♀)
egg cell before maturation: 60 μm diameter
secondary oocyte at ovulation: 120–150 μm diameter

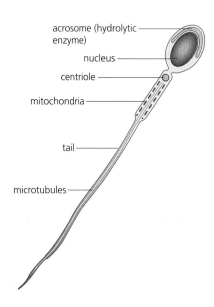

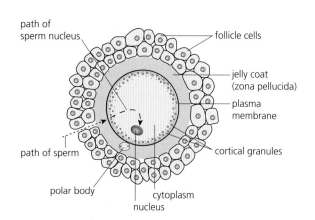

Steps leading to fertilisation

In mammals, fertilisation is internal and occurs in the upper part of the oviduct. The sperms are introduced into the female during sexual intercourse. The erect penis is placed in the vagina, and semen may be ejaculated (3–5 cm², in humans) close to the cervix. Typically, more than one hundred million sperms are deposited. The pH of the vagina is quite acid, but the alkaline secretion of the prostate gland, a component of the semen, helps to neutralise the acidity and provides an environment in which sperms can survive.

Waves of contractions in the muscular walls of the uterus and the oviducts assist in drawing semen from the cervix to the site of fertilisation. In this way, a few thousand of the sperms reach the upper uterus and swim up the oviducts. One or more of the few sperms that reach a secondary oocyte pass between the follicle cells surrounding the oocyte (Figure 3.25).

Next, the coat that surrounds the oocyte, made of glycoprotein and called the zona pellucida, has to be crossed. This is made possible by hydrolytic enzymes which are packaged in the tip of the head of the sperm, called the acrosome. In contact with the zona pellucida, these enzymes are released and digest a pathway for the sperm to the oocyte membrane. This process is part of the activation processes, called 'capacitation', in which sperms are prepared for fertilisation.

The head of the sperm, containing the male nucleus, is then able to fuse with the oocyte membrane. The nucleus enters the oocyte. As this happens, granules in the outer cytoplasm of the oocyte release their contents outside the oocyte by exocytosis. The result is that the oocyte plasma membrane cannot be crossed by another sperm.

As the sperm nucleus enters the oocyte, completion of meiosis II is triggered. The male and female haploid nuclei come together to form the diploid nucleus of the zygote. Fertilisation is completed.

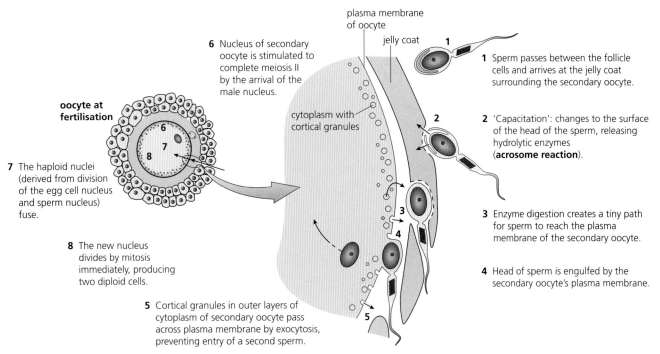

6 Nucleus of secondary oocyte is stimulated to complete meiosis II by the arrival of the male nucleus.

oocyte at fertilisation

7 The haploid nuclei (derived from division of the egg cell nucleus and sperm nucleus) fuse.

8 The new nucleus divides by mitosis immediately, producing two diploid cells.

plasma membrane of oocyte

jelly coat

cytoplasm with cortical granules

1 Sperm passes between the follicle cells and arrives at the jelly coat surrounding the secondary oocyte.

2 'Capacitation': changes to the surface of the head of the sperm, releasing hydrolytic enzymes (**acrosome reaction**).

3 Enzyme digestion creates a tiny path for sperm to reach the plasma membrane of the secondary oocyte.

4 Head of sperm is engulfed by the secondary oocyte's plasma membrane.

5 Cortical granules in outer layers of cytoplasm of secondary oocyte pass across plasma membrane by exocytosis, preventing entry of a second sperm.

Figure 3.25 The stages of fertilisation.

Fertilisation in flowering plants

Flowering plants contain their reproductive organs in the flower. **Flowers** are often hermaphrodite structures, carrying both male and female parts.

The **stamens** are the male parts of the flower, and consist of **anthers** (housing pollen grains containing the male gametes) and the **filament** (stalk).

The **carpels** are the female part of the flower. There may be one or many, free-standing or fused together. Each carpel consists of **ovary** (housing the ovules which each contain a female gamete), **stigma** (surface receiving pollen) and a connecting **style**.

Pollination and fertilisation

Pollination is the transfer of pollen from a mature anther to a receptive stigma. This pollen may come from the anthers of the same flower or flowers of the same plant, in which case, this is referred to as **self-pollination**. Alternatively, pollen may come from flowers on a different plant of the same species, which is referred to as **cross-pollination**.

Transfer of the pollen is usually by insects or by the wind, although in the flowers of certain species, running water or bird or bat visitors to the flowers may be the agents that carry out pollination. Insect pollinated flowers typically produce a sugar solution, called nectar, which attracts insects to visit the flower.

Fertilisation in flowering plants can occur only after an appropriate pollen grain has landed on the stigma, and germinated there. The pollen grain produces a pollen tube, which grows down between the cells of the style, and into the ovule through the micropyle (Figure 3.26). Incidentally, the pollen tube delivers two male nuclei. One of these male nuclei then fuses with the egg nucleus in the embryo sac, forming a diploid zygote. The other fuses with another nucleus which triggers formation of the food store for the developing embryo. This 'double fertilisation' is unique to flowering plants.

12 Explain the differences between pollination and fertilisation in the flowering plant.

Figure 3.26 Fertilisation
in a flowering plant.

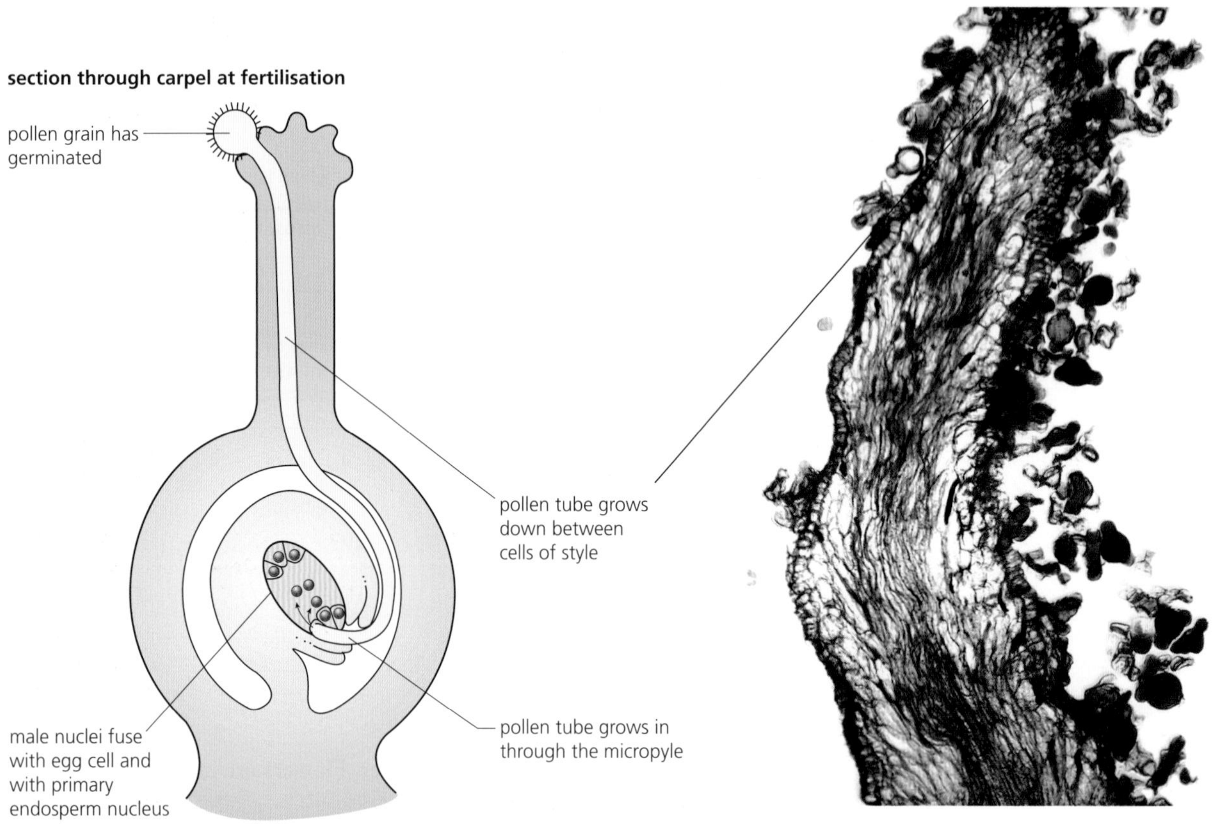

photomicrograph of LS through stigma, showing
germinating pollen grains (×100)

section through carpel at fertilisation

pollen grain has
germinated

pollen tube grows
down between
cells of style

male nuclei fuse
with egg cell and
with primary
endosperm nucleus

pollen tube grows in
through the micropyle

The significance of fertilisation in sexual reproduction

We have seen that gametes are formed by meiosis, and that the cells produced by meiosis differ
from each other for two reasons:

- **independent assortment** of maternal and paternal homologous chromosomes (Figure 3.23)
- **crossing over** of segments of individual maternal and internal homologous chromosomes.

The outcome is different combinations of genes in each gamete, which contributes to genetic
variety among the progeny. Variation lies at the heart of natural selection. We return to this
issue in Chapter 4.

In the event of fertilisation, a third cause of genetic variation arises, due to the new
combination of genes that comes from the random fusing of haploid gametes to form the diploid
zygote.

A* Extension 3.7:
Double fertilisation in
plants

3.4 Growth and development of the embryo

Growth is the permanent and irreversible increase in size that occurs in organisms with time;
development refers to the changes in shape, form and degree of complexity that accompany
growth of the organism. These processes are under the ultimate control of cell nuclei.

In the growth and development of a new organism from the zygote, the very first step is one of
continuous cell divisions to produce a tiny ball of cells. All these cells are capable of further
divisions, and they are known as embryonic stem cells.

A **stem cell** is a cell that has the capacity of repeated cell division while maintaining an undifferentiated state (self-renewal), and the subsequent capacity to differentiate into mature cell types (potency). Stem cells are the building blocks of life for they are able to proliferate and form cells that may develop into the range of mature cell types found in the organism. Stem cells are found in all multicellular organisms.

As the next stage of embryological development gets underway, more divisions of the cells formed from stem cells may occur, but eventually most cells lose the ability to divide. No-longer stem cells, they develop into one or other of the range of different tissues and organs that make up the organism, such as blood, nerves, liver, brain and many others. However, a very few cells within the tissues and organs of the developing organism retain many of the properties of embryonic stem cells, and these are called adult stem cells.

The differences between these two types of stem cell are summarised in Table 3.1.

Table 3.1 Differences between embryonic and adult stem cells.

Embryonic stem (ES) cells	Adult stem cells
Undifferentiated cells capable of continuing cell divisions and of developing into *almost all* the cell types of an adult organism (of which there are over 200 different ones).	Undifferentiated cells capable of cell divisions and of giving rise to a limited range of cells within a tissue type, e.g. blood stem cells give rise to red and white cells and platelets only.
Make up the bulk of the embryo as it commences development from the inner cell mass of the blastocyst (see Figure 3.27).	Occur in growing and adult body, within most organs, where they replace dead or damaged cells, e.g. in bone marrow, brain, liver etc.
Described as pluripotent stem cells – meaning having the potential to differentiate into 'very many but not quite all' cell types that make up the tissues and organs of the organism.	Described as multipotent stem cells – meaning capable of giving rise to a restricted range of cell types, i.e. those within the particular tissue or organ in which they are found.

Stem cells and society – a developing issue

You will be aware that the subject of stem cells is frequently mentioned in news headlines, today. Why this is – and what the properties, potential and associated ethical issues of stem cells may be – we shall discuss next. First, we will examine early human development to identify the practical sources of embryonic stem (ES) cells to which we will be referring.

Where do ES cells naturally occur?

A* Extension 3.8: The endometrium – venue for implantation

After fertilisation the zygote passes on down the oviduct towards the uterus, and cell division commences as it is swept along. The first few divisions form a stage known as cleavage, and they eventually result in a solid ball of cells being formed. The embryo does not increase in mass during cleavage; by the time the uterus is reached the individual cells are no larger than the fertilised egg cell from which they were formed (Figure 3.27).

Early in cleavage, at the 4–8 cell stage, the cells are especially versatile, for it has been shown that any one of these cells can develop into any and all the cells of the adult body. Consequently these cells are described as **totipotent stem cells** (having the potential to develop into *any* type of cell). So, for example, identical twins (or triplets, more rarely) arise when the cells of the embryo become completely separated during cleavage *at this very early stage.*

Division continues, and cells of the developing embryo organise themselves into a fluid-filled ball, known as the **blastocyst**. In humans, by day 7 the blastocyst consists of about one hundred cells. It now begins to embed in the wall of the uterus (called the endometrium). This process is known as **implantation.**

Next, some of the cells of the blastocyst become grouped together as the **inner cell mass.** It is these cells which eventually become the fetus proper. The **fetus** is the name for the human embryo from seven weeks after fertilisation, by which time it is recognisable as a tiny human being.

Once implanted, the embryo starts to receive nutrients directly from the endometrium of the uterus wall. At the same time, the other cells of the blastocyst begin to form the membranes that surround the fetus throughout its time in the uterus. Eventually, part of these membranes and

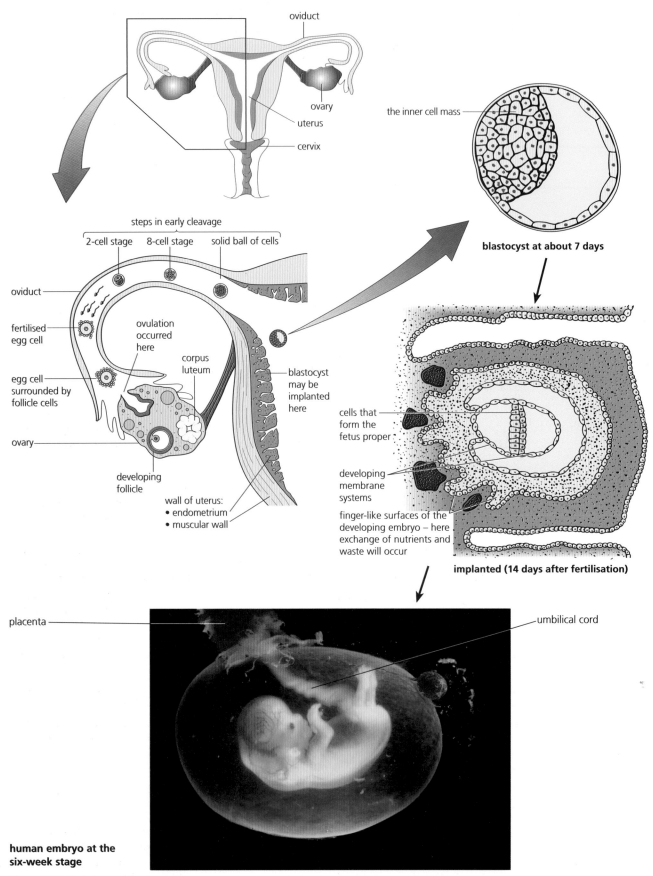

oviduct

ovary

uterus

cervix

the inner cell mass

blastocyst at about 7 days

steps in early cleavage

2-cell stage   8-cell stage   solid ball of cells

oviduct

fertilised egg cell

ovulation occurred here

corpus luteum

egg cell surrounded by follicle cells

ovary

developing follicle

wall of uterus:
• endometrium
• muscular wall

blastocyst may be implanted here

cells that form the fetus proper

developing membrane systems

finger-like surfaces of the developing embryo – here exchange of nutrients and waste will occur

implanted (14 days after fertilisation)

placenta

umbilical cord

human embryo at the six-week stage

Figure 3.27 Early human development.

13 Tabulate the differences between stem cells that are:
a totipotent
b pluripotent
c multipotent.

part of the endometrium form a new structure, the **placenta**. It is via this placenta that all exchange between fetal and maternal blood circulations will take place. The placenta is connected with the fetus by the **umbilical cord** – a vital life-line.

Are there roles for isolated stem cells in medical therapies?

We have seen the parts played by stem cells in formation of the new organism from the zygote. It is now appreciated that, if stem cells could be isolated in numbers and maintained in viable cell cultures, they might have uses in medical therapies to replace or repair damaged organs. To do this, once isolated, they require manipulation under reproducible conditions so that they:

- continue to proliferate, while held in a sterile cell culture environment, because relatively large volumes of tissue will be required
- are induced to differentiate into specific, desired cell types, like heart muscle
- survive in a patient's body into which they have been introduced, after differentiation
- integrate into the same tissue type in the patient's body
- continue to function correctly in the body for the remainder of the patient's life
- avoid triggering any harmful reactions among the cells and tissues of the patient's body.

Should these biotechnological and medical steps be completely mastered, then there are a range of medical conditions that arise from destruction or permanent damage to differentiated cells to which ES cell technologies could bring relief or offer a cure. Examples are listed in Table 3.2.

Table 3.2 Diseases that may eventually be treated by ES cell technology.

Multiple sclerosis (MS)	Multiple sclerosis arises when the fatty sheaths (myelin sheaths) that isolate, insulate and protect individual nerve fibres running to and from nerve cells (neurones) are attacked, (probably) by the body's immune system. This progressively destroys patches of fatty sheath cells, causing nervous communication to break down in areas of the body.
Parkinson's disease	Parkinson's disease arises from death of neurones (nerve cells) in part of the midbrain that controls subconscious muscle activities by means of release of a neurotransmitter substance called dopamine. Movement disorders result, with tremors in the hands, limb rigidity, slowness of movements and impaired balance.
Type 1 diabetes	Type 1 diabetes arises when the β-cells of the pancreas are destroyed by the body's immune system, and a severe lack of insulin results. Insulin (a hormone) maintains the blood glucose concentration at about $90\,mg\,100\,cm^{-3}$. In diabetics, the level of blood glucose becomes erratic and generally permanently raised. Glucose is regularly excreted in the urine.
Duchenne muscular dystrophy	Duchenne muscular dystrophy results from a deficiency in a muscle protein, dystrophin, normally produced by muscle fibres. These become progressively replaced by non-contracting fibres. Typically, boys have walking difficulties from 1–3 years, and at 8–11 years become unable to walk. From late teens, muscular weakness may start to be life-threatening.
Spinal cord injury	Spinal cord injury occurs when a blow to the spine fractures or dislocates vertebrae, and some, many or most axons (nerve fibres) connecting parts of the body and the brain are destroyed. At worst, total lack of sensory and motor function below the level of injury results, but incomplete injuries leave variable impairment and varying degree of recovery.
Burns	Burns, due to heat or other cause, may be mild (first degree) to severe (third degree). The latter destroy the deepest layers of skin, and may need skin grafts to achieve a full recovery.
Brain damage	Brain damage may be due to a stroke, caused by a clot or haemorrhage and resulting in loss of brain function as neurones (nerve cells), deprived of oxygen and glucose, quickly die.
Cardiac muscle damage	Cardiac muscle damage occurs due to myocardial infarction, caused by major interruption to the blood supply to areas of cardiac muscle sufficient to cause the death of muscle fibres.

How stem cells may be obtained

There are a number of ways that stem cells may be obtained, provided that the community approves the technology and purposes of the research.

1 **Embryonic stem cells may be obtained from 'spare embryos'** produced by infertility clinics while treating infertile couples. Either the male or female or both partners may be infertile, due to a number of different causes. In some cases, infertility may be overcome by the process of fertilisation of eggs outside the body, known as *in vitro* fertilisation (IVF). The process is

outlined in Figure 3.28. ES cells may be obtained from the inner cell mass (Figure 3.27) in embryos not to be used in fertility treatments, where this is allowed and agreed. Today, the procedure is regarded as a routine one, although using the numerous spare embryos for ES cell research remains controversial, as we shall see. The chief objection is that the embryo's 'life' is destroyed in the process.

Figure 3.28 *In vitro* fertilisation – the process.

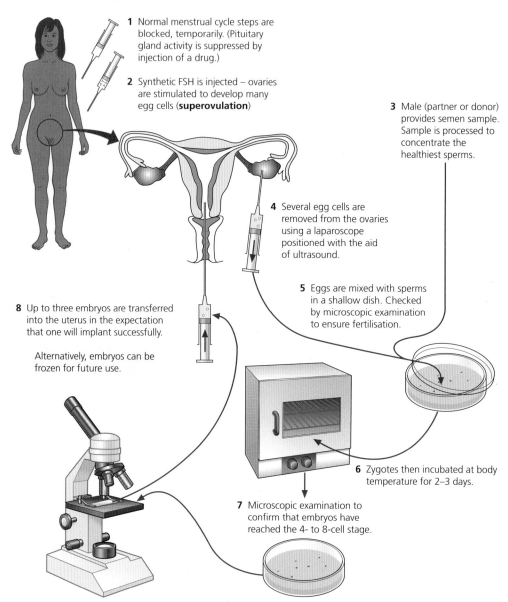

1 Normal menstrual cycle steps are blocked, temporarily. (Pituitary gland activity is suppressed by injection of a drug.)

2 Synthetic FSH is injected – ovaries are stimulated to develop many egg cells (**superovulation**)

3 Male (partner or donor) provides semen sample. Sample is processed to concentrate the healthiest sperms.

4 Several egg cells are removed from the ovaries using a laparoscope positioned with the aid of ultrasound.

5 Eggs are mixed with sperms in a shallow dish. Checked by microscopic examination to ensure fertilisation.

8 Up to three embryos are transferred into the uterus in the expectation that one will implant successfully.

Alternatively, embryos can be frozen for future use.

6 Zygotes then incubated at body temperature for 2–3 days.

7 Microscopic examination to confirm that embryos have reached the 4- to 8-cell stage.

2 An alternative source of ES cells is **from a single cell removed from an embryo at the eight-cell stage** of cleavage (Figure 3.27). For some time now it has been known that a single cell from this group of eight can safely be removed for the purpose of pre-implantation genetic diagnosis (PGD – carried out in cases where partners are seeking to avoid an embryo carrying a specific genetic defect) because the rest of the embryo is not destroyed or harmed. The remaining seven cells develop into a normal blastocyst, and, if implanted, have the ability to grow into a healthy offspring.

So, this procedure can be adapted to allow PGD, the growth of a culture of ES cells for possible medical therapies, *and* the implantation of a healthy blastocyst into the uterus. No loss of 'life' will have occurred (Figure 3.29).

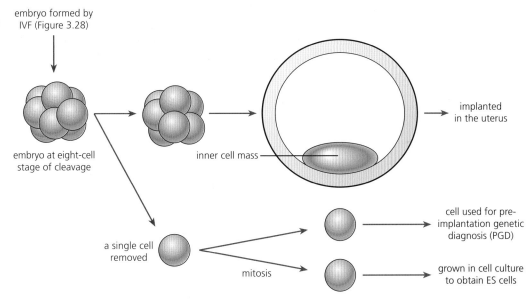

Figure 3.29 Obtaining ES cells without loss of the blastocyst.

embryo formed by IVF (Figure 3.28)

embryo at eight-cell stage of cleavage

inner cell mass

implanted in the uterus

a single cell removed

mitosis

cell used for pre-implantation genetic diagnosis (PGD)

grown in cell culture to obtain ES cells

3 Another source of ES cells is the blood that can be **extracted from the umbilical cord** at the time of birth. 'Cord blood', as it is referred to, is normally disposed of, yet it has been shown to contain cells that are pluripotent and so indistinguishable from the ES cells obtained as described above. Samples of cord blood (typically of 40–100 cm³) are collected, the stem cells harvested, and then 'multiplied' by sterile cell culture technique to yield sufficient ES cells for practical purposes. Since one hundred million babies are born each year, this source should surely grow to be significant.

4 Finally, **sources of adult stem cells** are sought for possible medical therapies, too. They have been identified in many organs and tissues including the brain, bone marrow, skin and liver, although they are present there in tiny quantities and in a non-dividing state. These stem cells are naturally activated in and by the event of damage or disease of the organ in which they occur.

    Actually, the stem cells that generate blood cells (known as haematopoietic cells) are obtained from bone marrow, and are already used in treatments. When patients are to have high doses of chemotherapy and or radiotherapy then bone marrow transplants are withdrawn from bones in their body like the femur, hip bones, ribs or sternum and held for later return. This is necessary because anti-cancer treatments like these are indiscriminate and kill very many of any actively dividing cells present (not just cancer cells).

    You may also have read about the use of bone marrow transplants to treat leukaemia. There are sometimes very harrowing reports of frantic searches for donors with compatible cells so that treatment can be given to a child with leukaemia before the disease overtakes the patient.

14 What do we mean by *in vitro* and *in vivo*?

A* Extension 3.9: Stem cells from testicles – a recent development

Activity 3.11: Wider reading – 'New cells for old'

Regulation of ES cell research, and the ethical issues it generates

The first 'test tube baby' (resulting from IVF treatment) was born in 1978, and immediately triggered a national debate among scientists and the medical profession, and among organisations representing parents and patients. The debate was joined by some philosophers, by religious organisations and political leaders. The government responded by appointing an advisory committee in 1982, under the chairmanship of the philosopher Mary Warnock.

The Warnock Report, as it is known, had a profound, thoughtful and widely respected impact on the controversial and highly sensitive issues that IVF treatments generate.

The main action outcome was the setting up the Human Fertilisation and Embryology Authority (HFEA) in 1991, as required by an Act of Parliament past in the preceding year. You can read a resumé of the roles and duties of the HFEA in Table 3.3.

Table 3.3 The Human Fertilisation and Embryology Authority (HFEA).

The roles	1 To license and monitor fertility clinics that carry out IVF and donor insemination, centres undertaking human embryo research, and all storage of human gametes and embryos. 2 To produce a Code of Practice on the proper conduct of HFEA-licensed activities, as guidelines to clinics. 3 To review information on HFEA-licensed activities, monitor any subsequent developments, and advise the Secretary of State for Health on developments.
Licensing responsibilities	1 To issue treatment, storage and research licenses to centres, and to renew them on an annual basis. 2 To define and promote good clinical practices and high ethical standards, and ensure clinics comply with requirements of the HFE Act and the European Tissue and Cells Directive.

The ethical issues this area of human reproduction technology generates centre on:

- IVF as a medical practice, for infertility is a personal problem that generates stress and unhappiness in those affected
- the obtaining of ES cells from embryos, because of widely differing views about the status of early embryos.

Tables 3.4 and 3.5 are summaries of divergent arguments.

Table 3.4 Ethical issues raised by IVF.

Favourable arguments	Critical arguments
For some otherwise childless couples, desired parenthood may be achieved.	Allows infertility due to inherited defects to be passed on (unwittingly) to the offspring who may then experience the same problem in adulthood.
Allows men and women surviving cancer treatments the possibility of having children later, using gametes harvested prior to radiation or chemotherapy treatments.	Excess embryos are produced to ensure success, and so an embryologist has to select some new embryo(s) to live, and to allow the later destruction of other, potential human lives.
Permits screening/selection of embryos before implantation stage, to avoid an inherited disease.	Multiple pregnancies have been a common outcome, sometimes producing triplets, quads or sextuplets, leading to increased risk to the mother's health, and to risks of premature birth and babies at risk of cerebral palsy.
If IVF treatment were to be banned, the State may be interfering in the lives of individuals with medical problems that otherwise can be cured.	Infertility, it can be argued, is not always strictly a health problem; it may have arisen in older couples who choose to delay having a family (a lifestyle issue).
Offspring produced by IVF are much longed-for children who will more certainly be loved and cared for.	There is an excess of unwanted children, cared for in orphanages or in foster homes. These children may have benefited from adoption by couples, childless or otherwise, keen to be caring 'parents'.

What other factors do you feel should be added?

15 Outline key points that might be put by a genetic counsellor who felt that some alternative way to establish a family was more appropriate for a particular childless couple seeking IVF treatment.

Table 3.5 The ethics of obtaining and using ES cells.

Favourable arguments	Critical arguments
Respect due to an early-stage embryo as a human being increases significantly as it develops, and needs to be weighed against the potential future benefits of ES cell research.	An embryo should be accorded full human status from the moment of its creation – held by some people to occur as soon as a zygote begins to divide.
The ultimate fate of all *spare* embryos produced in IVF treatments is to be destroyed.	All human life is sacred at whatever level or stage of development, and no human life of this sort should be taken in these circumstances.
It is now possible to obtain ES cells from an embryo at the eight-cell stage without causing the death of the embryo (Figure 3.2).	It is an important overriding principle for some people that humans should not tamper with 'nature' in a deliberate way.
It is possible to regulate and control ES cell research to serve: ■ advances in understanding and treating infertility and the causes of miscarriages ■ future opportunities of treating and possibly curing conditions such as those listed in Table 3.2 above ■ investigation of the causes of congenital diseases and the development of methods of detecting gene or chromosome abnormalities.	Stem cell research is a costly technology, mostly beneficial to the lives of a very limited number of people of developed nations, whereas if much of the funds used were to be diverted to solve more basic problems of health and nutrition (e.g. clean water supply, effective contraception) of the poor worldwide and to many in the less developed countries, the money would benefit vastly more humans.

What other factors do you feel should be added?

Table 3.5 The ethics of obtaining and using ES cells.

HSW 3.3: Criterion 10 – Ethical issues in the treatment of humans

Activity 3.12: A website on human reproduction technologies and the applications of ES cells

---

■ **How Science Works 3.4: Criterion 4 – Demonstration of totipotency using plant tissue culture techniques**

It is not possible for you to work with animal embryos for many reasons, but laboratory investigations are possible using plant tissues.

Using the internet, research 'plant tissue culture' for sites that give practical guidance.

You need to gather the necessary information to plan your own demonstration of totipotency, using the facilities available to you in laboratories you have access to for Biology practicals and investigations.

Discuss your plans with your tutor to arrive at the best procedures to adopt. Sources that will help include:

Science and Plants for Schools (SAPS): www-saps.plantsci.cam.ac.uk

Kitchen Culture Kits Inc.: www.home.turbonet.com/kitchenculture/sivbposter.htm

Genes and the control of development

Gene expression

In Chapter 2, we discussed the role of DNA of the chromosomes in instructing the cell to make specific proteins. Within these extremely long molecules, the relatively short lengths of DNA that code for single proteins are called **genes**. Proteins are very variable in size (and so, too, are genes), but most proteins contain several hundred amino acids condensed together in a linear series. The prelude to production of a protein is the transcription of the code of a gene into mRNA (page 77).

Some of the genes in a cell's chromosomes are actively transcribed throughout the life of the cell, but others are activated (we say they are **expressed**) only at a particular stage in the life of the cell, or when the substance they act on (their substrate molecule) is present. Very many of our genes have to be deliberately activated, as required. Obviously the genes concerned with development of the organism from the zygote are some of the first to be expressed, but most probably for a limited period only.

16 What do the terms *genome* and *gene* mean?

How are individual genes switched on?

One activating mechanism that has been found to operate in bacteria (prokaryotes) is the **lactose operon** mechanism (Figure 3.30). This is worth looking into because it gives us an idea of how the issue of genes being switched on was resolved in one organism, at least.

This mechanism consists of a **regulator gene** and an **operator gene**, close to the genes that are regulated (known as a **structural gene** – in this case, coding for a lactose-metabolising enzyme). The regulator gene codes for a repressor protein which binds to the operator gene site. The repressor protein prevents transcription of the structural gene. However, if lactose is present (for example, it might become available in the medium in which the bacteria are growing), then the lactose molecule reacts with the regulator protein, preventing it from binding with the operator gene. As a result, the lactose-metabolising-enzyme gene is transcribed, and lactose is metabolised. Once all the lactose has been used up, the repressor molecule blocks transcription again.

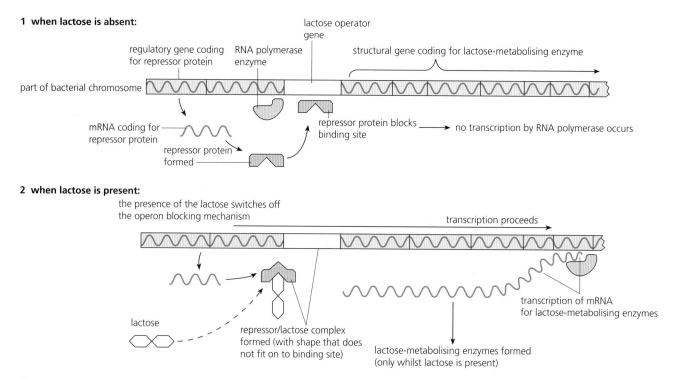

Figure 3.30 'Operon' gene regulation.

Of course, this basic type of mechanism occurs in prokaryotes (bacteria and cyanobacteria) only. In eukaryotes, the regulator-gene mechanisms are more complicated. (They are briefly summarised in Figure 3.31.) However, the principle is the same, in that transcription of many structural genes is regulated according to the needs of the cell.

Figure 3.31 Control of gene expression in eukaryotes – a summary.

In the nucleus

1 RNA polymerase requires particular transcriptional initiation factors to function – this is the chief way that eukaryotic gene expression is controlled.

2 mRNA processing selectively removes nonsense DNA called introns. Here, alternative RNA splicing generates different mRNA molecules.

3 Movement of mRNA through pores in nuclear membrane is an active, selective process – only mRNA that passes out to ribosomes is transcribed into proteins (gene product).

mRNA produced by transcription – a copy of the coded information of a gene

many genes contain short lengths of 'nonsense' (sequences of bases additional to coded information of the gene) called introns

introns are 'edited' out of the mRNA by the action of enzymes present in the nucleus

introns

exons

what remains, known as exons, are joined up to form mature mRNA

mature mRNA passes out of the nucleus via pores in the nuclear membrane

5′

3′

pore in nuclear membrane

to ribosomes in the cytoplasm

In the cytoplasm

4 Post-transcriptional modification may change the gene product selectively, and so influence gene expression.

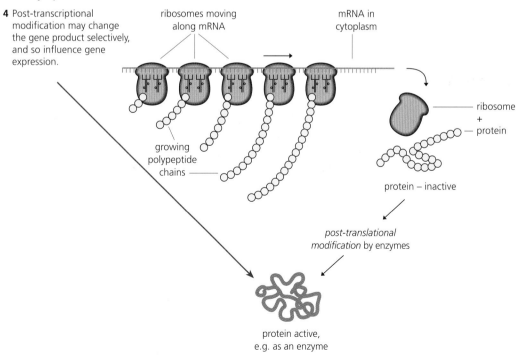

ribosomes moving along mRNA

mRNA in cytoplasm

growing polypeptide chains

ribosome + protein

protein – inactive

post-translational modification by enzymes

protein active, e.g. as an enzyme

■ Extension: How development is controlled

We know that every nucleus holds a complete copy of the genome of the whole organism. In human embryos up to the eight-cell stage of embryological development the cells are totipotent. (Cells of that embryo can be detached and form another complete embryo.) But after this stage in development most cells of the embryo become progressively and irreversibly committed to development into a particular tissue (and eventually, a particular organ) of the adult organism. At this point onwards, in differentiating cells some of the genes relating to the structure and function of other types of cell of the organism will remain 'switched off'.

The challenge is to understand how both the environment within these early cells and the cell's positions within the tiny embryo influence the path of development.

It appears that cells now manufacture tissue-specific proteins in a sequence that directs maturation into particular tissues and organs. How this 'determinism' is controlled is outside the scope of this book. Suffice to say the phenomenon is studied in detail and understood in part.

Interactions of genes – polygenes and environmental effects

We began the story of genetics in Chapter 2 with an investigation of the inheritance of height in the garden pea (page 84), where one gene with two alleles gave 'tall' or 'dwarf' plants. This clear-cut difference in an inherited characteristic is an example of **discontinuous variation** in that there is no intermediate form, and no overlap between the two phenotypes.

In fact, very few characteristics of organisms are controlled by a single gene. Mostly, characteristics of organisms are controlled by a number of genes. Groups of genes which together determine a characteristic are called **polygenes**.

Polygenic inheritance is the inheritance of phenotypes that are determined by the collective effect of several genes. These genes interact with one another.

The genes that make up a polygene are often (but not necessarily always) located on different chromosomes. Any one of these genes makes a limited impact on the phenotype, but the combined effect of all the alleles of the polygene is to produce infinite variety among the offspring where large numbers are produced.

There follows discussion of five examples of gene and environment interactions.

1 Human height

Many features of humans are controlled by polygenes, including body height (and weight). The graph of the variation in the heights of a population of 400 people in Figure 3.32 shows continuous variation in height, between the shortest at 160 cm, and the tallest at 186 cm. The mean height is 173 cm. Human height is controlled by several different genes.

Figure 3.32 Human height as a case of polygenic inheritance.

Human height is determined genetically by interactions of the alleles of several genes, probably located at loci on different chromosomes.

Variation in the height of adult humans
The results cluster around a mean value and show a normal distribution. For the purpose of the graph, the heights are collected into arbitrary groups, each of a height range of 2 cm.

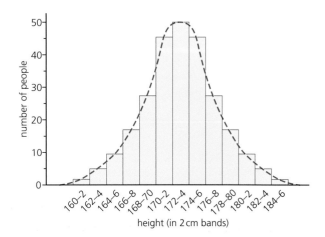

number of people

height (in 2 cm bands)

In humans, many genetically determined features may be influenced by environmental factors, too, and body height is one. Here, the amount and quantity of nutrients received at critical stages of body growth and development are equally likely to influence the phenotype as the polygenes that control height genetically. In this case the relative contributions of genes and environment are very difficult to determine.

2 Skin colour in humans

The colour of human skin depends primarily on a pigment called **melanin**, which is produced by special cells called melanocytes, present in the epidermis of the skin (Figure 3.34).

Skin colour is under polygenetic control. It seems that at least three or four or more genes control the various aspects of melanin production and deposition in skin cells. The outcome of

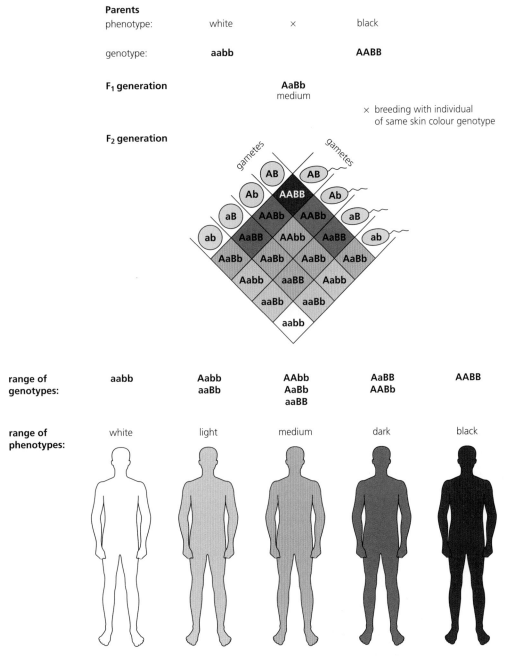

Figure 3.33 Human skin colour as a characteristic controlled by two independent genes – an illustration of polygenic inheritance.

17 Derive the ratio of phenotypes produced in the F$_2$ generation shown in Figure 3.33.

the interactions of the alleles of these genes is an almost continuous distribution of skin colours among humans from very pale (presence of no alleles coding for melanin production) to very dark brown (all 'skin colour' alleles coding for melanin production). Figure 3.33 is a simplified illustration of polygenic inheritance of human skin colour, for it involves only two of the three or four independent genes involved. This is because dealing with all four genes creates a hopelessly unwieldy genetic diagram. However, the principle can be demonstrated clearly enough using just two genes, for even here we see more or less continuous variation in the range of phenotypes.

How does human skin colour form?

The structure of the skin is shown in Figure 3.34. Melanocyte cells manufacture melanin in tiny organelles called melanosomes, by the catalytic activity of an enzyme, tyrinosinase, acting upon the amino acid tyrosine. After manufacture, melanosomes are transferred to the outermost cells of the dermis. Each melanocyte serves about 40 skin cells in this way. The overall picture is further complicated by the fact that two forms of melanin are produced, one red–yellow in colour, and the other brown–black. The former is common to lighter skins, the latter to darker skins, but the ratio of these two pigments is apparently less significant than the total amount of melanin formed. Also important is the ultimate size and distribution of the melanocytes. In a more pigmented skin, the melanosomes are larger and dispersed more widely in skin cells.

How may environmental conditions influence skin colour?

Enhanced exposure to UV light increases the enzyme activity within the melanosomes, and the melanocytes are stimulated to transfer their melanosomes to skin cells more quickly (Figure 3.34). Here they are seen to collect around the nucleus and become dispersed in the cytoplasm. As a result, skin darkens, and skin cells form a more effective barrier against penetration of UV light.

■ Extension: Albino condition

We have already discussed the inheritance of the albino condition (pages 88–89). Albinos have a gene mutation that blocks the formation of the enzyme tyrinosinase altogether, and they are unable to form melanin anywhere in their bodies. For albinos, their hair and skin colour – or the lack of it – is entirely under genetic control.

18 What is meant by the term *mutant*?

3 Animal hair colour

The fur of mammals consists of densely packed hairs, of the type shown in the skin section in Figure 3.34. Since the hair shaft is a product of the germinal layer of cells (Malpighian layer) that forms the outer cells of the skin (granular and cornified layers), it too may be impregnated by melanin under the influence of alleles of the genes determining skin colour in that particular species. Figure 3.35 shows some mammals with distinctive fur colouring.

Frequently, fur is moulted at one or more times in the year, so the thickness of this body insulation layer varies. Seasonal responses like these are typically a result of day-length changes detected in the animal's brain via signals from the eyes (rather than as a result of the detection of changing temperatures). Remember, at higher latitudes, day-length changes between summer and winter are very marked.

In certain mammals conditions in the environment also influence coat colour. For example, in just a few mammals, the colour of the fur changes seasonally. Examples are listed in Table 3.6. We assume that the seasonal environmental changes detected in the brain lead to selective modulation of the impact of individual alleles determining colour in the hair formed in winter and in summer periods, in these species.

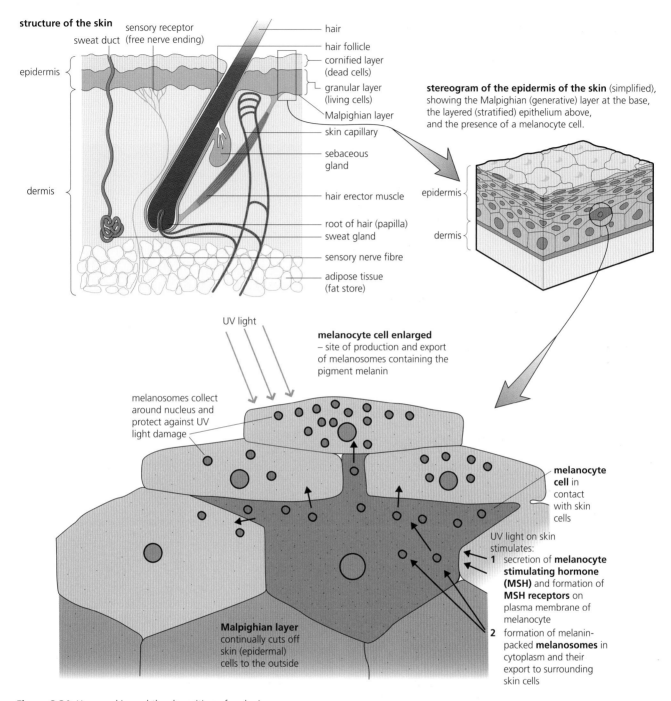

Figure 3.34 Human skin and the deposition of melanin.

Table 3.6 Examples of mammals with changing coat colours.

	Winter coat colour	Summer coat colour
Arctic fox (Figure 3.35) (*Alopex lagopus*)	white	brownish-grey
Arctic hare (*Lapus arcticus*)	white (with black at tips of ears)	grey-brown above, white below
Stoat (*Mustela erminea*)	may be white according to location	chestnut brown

Certain other mammals have a mutation in the gene coding for the enzyme tyrosinase, the enzyme catalysing the formation of melanin. This mutated enzyme is heat-sensitive; it fails to work at normal body temperatures, but becomes highly active in cooler areas of the skin. This results in dark colouration in the coolest parts of the body.

Look at the photographs of the Siamese cat and Himalayan rabbit in Figure 3.35. In both these animals, marked and persistent changes in ambient temperature affect the whole body colour. Cold weather intensifies the markings, while persistent hot weather causes them to fade.

Figure 3.35 Summer and winter coat colour in the Arctic fox, and distinctive coat colouration in the Siamese cat and Himalayan rabbit.

4 Cancers

There are very many different forms of cancer, affecting different tissues of the body; cancer is not thought of as a single disease. However, in all cancers, cells divide by mitosis repeatedly, without control or regulation, forming an irregular mass of cells called a **tumour**. Tumour cells then emit signals promoting the development of new blood vessels to deliver oxygen and nutrients, all at the expense of the surrounding, healthy tissues. Sometimes tumour cells break away and are carried to other parts of the body, forming a secondary tumour (a process called metastasis). Unchecked, cancerous cells ultimately take over the body, leading to malfunction and death.

So, cancer arises when the cell cycle (Figure 3.14, page 109) operates without its normal controls. *Look back at the stages of the cell cycle now.*

Note that the cell cycle consists of distinct phases, represented in shorthand as $G_1 \rightarrow S \rightarrow G_2 \rightarrow M \rightarrow C$.

How is the cell cycle controlled in a healthy cell?

The cell cycle is regulated by a molecular control system which is the subject of current research. The key points of this system are listed below, and are best understood by studying them in conjunction with Figure 3.36.

- In the cell cycle there are key **checkpoints** where signals operate; these are 'stop' points which have to be overridden.
- Three checkpoints are recognised – at G_1, G_2 and in M. All are important.
- At the G_2 checkpoint, if the 'go-ahead' signal is received here, the cell goes through to M → C, for example.
- if a 'no-go' signal is received at the checkpoints, the cell passes into a non-dividing state, referred to as G_0. Incidentally, most cells in the *adult* human body are in a G_0 phase, and some are unable to be re-activated (for example, muscle cells, nerve cells), whereas others (for example, liver cells) may be re-activated by external growth factors released by tissue damage, for example.
- The molecular control signal substances in the cytoplasm of cells are proteins known as **kinases** and **cyclins.**
- Kinases are enzymes that either activate or inactivate other proteins by a phosphorylation reaction (the addition of phosphate). Kinases are present in the cytoplasm all the time, though sometimes in an inactive state.
- Kinases are activated by a cyclin, so they are referred to as cyclin-dependent kinases (Cdks).
- Cyclin concentrations in the cytoplasm constantly change. As the concentrations of cyclins increase, they combine with Cdk molecules to form a complex which functions as a mitosis-promoting factor (MPF).
- As MPF accumulates, it triggers chromosome condensation, fragmentation of the nuclear membrane, and finally spindle formation – that is, mitosis is switched on.
- By anaphase of mitosis, destruction of cyclins commences (but Cdks persist in the cytoplasm).
- Certain external factors also operate upon the cell, either triggering rises in cyclin concentration or switching on the destruction of cyclin.

Activity 3.13: Wider reading – 'The cell cycle and mitosis'

Figure 3.36 The molecular control system of the cell cycle.

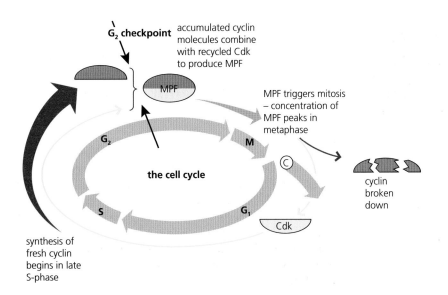

G_2 **checkpoint**

accumulated cyclin molecules combine with recycled Cdk to produce MPF

MPF

MPF triggers mitosis – concentration of MPF peaks in metaphase

G_2

M

the cell cycle

C

cyclin broken down

S

G_1

Cdk

synthesis of fresh cyclin begins in late S-phase

What may happen in cells to trigger tumour formation, leading to a cancer?

Cancer is principally caused by damage to DNA. Such damage may be due to physical factors such as UV light, X-rays or gamma rays, or by chemicals we call carcinogens. The latter may also come from the environment, including components of the tar in tobacco smoke, or they may be molecules manufactured within the organism.

'Mistakes' or mutations of different types build up in the DNA in body cells. The accumulation of mistakes with time explains why the majority of cancers arise in older people.

Different types of mutation can occur, and in cells of different body tissues, and hence cancer is not one single disease. Whatever the cause, two types of genes play a part in triggering cancer, if they mutate:

1 **oncogenes** – the genes that code for the proteins that stimulate the cell cycle. Mutations in an oncogene may result in excessive cell division. In this case, the cells can go on dividing indefinitely, and if supplied with sufficient nutrients they are 'immortal'.
2 **tumour-suppressing genes** that code for proteins that stop a cell cycle. Mutations may inactivate this type of gene, such as the one that codes for a protein known as p53, which stops the copying of damaged DNA. (This gene and its p53 protein have been described as 'the guardians of the genome'.) The undamaged gene and its p53 protein work by halting the copying of faulty DNA, so that other enzymes may then repair and correct the fault – cancer is avoided.

Another cause of certain cancers is a specific virus infection. Such a virus can trigger cancer when its DNA becomes integrated with that of the cell it has infected. A mutated oncogene may be added, or an 'unstable' proto-oncogene that rapidly mutates. Other virus DNA may code for a protein that inactivates p53 protein present in the host cell.

Another set of factors are inherited, in that the members of some families are more likely to suffer from certain cancers than others. In effect, certain cancers run in families. Examples include a form of colorectal cancer, and also breast cancer. In the latter case, two genes (known as BRCA1 and BRCA2) may be inherited, each with one mutant allele, and in these cases there is a 60% probability that the healthy allele will also mutate at some stage, before the person reaches the age of 50 years, thereby triggering breast cancer. Both BRCA1 and BRCA2 are natural tumour-suppressing genes when present in the homozygous state.

Finally, Table 3.7 is a listing of the six profound abnormalities exhibited by cancer cells that overrun healthy tissues and fatally spread throughout the body.

19 Describe how the behaviour of cancer cells differs from that of normal cells.

Table 3.7 Hallmarks of cancers.

Cells amplify external growth factors or generate their own.
Cells have lost their natural cell-cycle suppression system.
Cell suicide system, a natural component of healthy cells, is disabled or overridden.
Cells evade the natural restriction system limiting the number of times division can occur.
Cells emit signals that develop new and expanding nutrient supply channels.
Cells overrule forces restricting movement – allowing survival among other body tissues.

5 The strange case of monoamine oxidase (MAO A)

Control and co-ordination of the body of a mammal is by the action of the nervous system. This is built from specialised nerve cells called neurones, which transmit nerve impulses along extended cell structures called fibres. At the ends of individual fibres are tiny gaps called synapses. While transmission of an impulse along a fibre is by temporary reversal of potential difference along the fibre membrane (sometimes referred to as electrical transmission), transmission at the synapses is by diffusion of special chemicals. These substances are known as neurotransmitter substances. Several different neurotransmitters exist, some operating throughout the nervous system, others restricted to particular parts, such as the brain.

DL
www
A* Extension 3.10:
Why are males
vulnerable to abnormal
MAO A activity?

Not surprisingly, there are present in the body, specialised enzymes that regulate the levels of various transmitter substances. Monoamine oxidase (MAO A) is one such; it degrades a range of amine-containing neurotransmitters (including dopamine). The production of too little or too much MAO A may be associated with neurological disorders, possibly leading in some cases to violent criminal behaviour. Some violent criminals have been found to have low MAO A activity. These people have mostly been males.

The allele coding for the production of the enzyme MAO A is located on the X chromosome (a sex chromosome). It is mutation of this allele that may lead to individuals exhibiting violent behaviour. However, the symptoms are largely restricted to young people and adults who have already experienced maltreatments as children. So, while the potential for this condition is present at birth, it is observed only later, in certain individuals who have been exposed to particularly disturbing environmental conditions. This is a rather complex example of the interplay between environmental factors and genotype, expressed in the phenotype.

DL www 3 End-of-topic test

(full End-of-topic tests are available on the DL Student website)

1 What are the specific roles of each the following organelles when deployed in the synthesis of a specific protein destined for discharged from the cell?
a nucleus
b RER
c Golgi apparatus
d vesicle (8)

2 The following is an account of an investigation of development of a multicellular organism. The experiment was conducted before the nature of stem cells was understood. Read the text and then fill in the blanks, using the most appropriate term from this list: genome; *in vitro*; stem cells; embryo; medium; totipotent; genes; zygote; nutrients; expressed

How do we know that all cells in an adult, multicellular organism retain the same genes throughout life, identical with those of the single cell (_____) from which that individual originally grew, as a product of sexual reproduction?

This was first demonstrated by F.C. Steward and co-workers, at Cornell University, USA, in the 1950s, working with plants. This team found that a mature cell (it was a parenchyma cell obtained from carrot root tissue) could be dislodged, isolated, and then cultured in an appropriate liquid nutrient medium. To be successful, this _____ contained not only essential _____, but also certain plant growth substances.

Each isolated cell not only survived, but started to divide and form an aggregate of living cells. These cells then formed an embryonic plant with a root (radicle), stem (plumule), and embryonic leaves (cotyledons). This occurred when cultured in glass apparatus (_____), when supplied with nutrients and growth factors. Further, this isolated _____ then developed into an entire, mature plant, also *in vitro*.

The conclusion of this experiment was that the cells of an organism, from first formed till last, had the same _____. During the life of an organism, different _____ of the total complement are activated (we say they are _____), and responsible for development whilst other genes are temporarily inactive. In effect, genes are selectively active as the zygote develops into an embryo, and then the embryo grows into an adult.

Today we would say that Steward and his team had created _____ from mature, fully differentiated cells. Further, these stem cells were _____ for they gave rise to all the mature cells of the organism. (10)

3 a Place the following stages of the cell cycle in their correct sequence, commencing with division of the cell itself:
interphase-G_2 mitosis interphase-S cytokinesis interphase-G_1
b Identify the following aspects of the process by which the DNA content is doubled:
  i stage in the cell cycle
  ii nature of process in the nucleus
  iii chief enzyme responsible.
c In control of the cell cycle, explain the significance of the following:
  i check points
  ii role of cyclins.
d What is the significance of mitosis in:
  i the growth of a multicellular organism
  ii the process of asexual reproduction? (12)

4 Male and female mammalian gametes are specialised cells, but are different in formation, structure and behaviour. Complete the table of possible features of sperms and eggs by placing a tick (✓) or cross (✗) as appropriate. (8)

Feature	Sperm	Egg
adapted for swimming motion		
has a haploid nucleus		
carries an organelle packed with hydrolytic enzymes		
cytoplasm contains cortical granules		
has region packed with mitochondria		
surrounded by jelly coat and follicle cells		
largest dimension less than 10 m		
largest dimension more than 10 m		

5 Explain the difference between a tissue, an organ and an organ system by reference to a specific example. (9)

6 a Define the terms 'genotype' and 'phenotype'.
b By reference to particular characteristics of named species, explain what you understand by 'continuous variation' and 'discontinuous variation'.
c Outline how continuous variations may arise. (12)

7 a Draw a labelled diagram showing the distinctive structural features of a bacterium.
b Detail how a mammalian liver cell differs from a bacterium in terms of:
   i size **ii** genetic material **iii** organelles (12)

8 Our knowledge of ultrastructure of animal cells is based on interpretation of images obtained from thin sections of tissue examined by electron microscopy.
a Explain why cells from an animal tissue require the following treatments, prior to examination by the transmission electron microscope:
   i killing and fixing **ii** thin sectioning
   iii staining **iv** dehydrating.
b The diagram below is of a mammalian liver cell. Identify the structures labelled **A–E**.

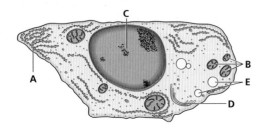

c The magnification of the above image of a liver cell is ×1800. What is the length of this cell in μm (microns)? Show your workings. (15)

9 a During prophase of mitosis in an animal cell, what changes occur to the following structures:
   i chromatin **ii** nuclear membrane **iii** centrioles?
b How do the structure, position and behaviour of the chromosomes differ during metaphase and anaphase of mitosis?
c Identify the particular tissue of a named organism that you would select to observe stages in mitosis, and state why.
d Outline the steps by which the tissue selected in part **c** above is prepared for microscopic examination of stages in mitosis.
e When preparations showing stages in mitosis are examined under the microscope, the numbers of cells observed in the four phases varies consistently. For example, most are found to be at prophase and fewest at anaphase. How can this difference be best explained? (20)

10 a What do you understand by the term 'genetic variation'?
b Explain how independent assortment arises in meiosis.
c When does crossing over occur in the process of meiosis and how does it contribute to genetic variation? (10)

11 a Define 'fertilisation'.
b The events leading to fertilisation in flowering plants commence after pollen has reached the stigma. Answer the questions below concerning the steps involved in fertilisation of the egg cell in an ovule.
   i How does the germinating pollen grain deliver its male nuclei to the embryo sac?
   ii There are two male nuclei delivered to the embryo sac. Describe the fate of both nuclei, separately, together with the specific products of this double fertilisation event.
c The process of fertilisation in the mammal commences with the arrival of one or more sperms at the secondary oocyte. Answer the questions below concerning the steps involved in fertilisation of the oocyte.
   i Where does fertilisation occur within the female?
   ii How is the jelly coat outside the oocyte penetrated?
   iii Once the head of a sperm has penetrated the oocyte plasma membrane, how is entry of a further sperm nucleus prevented?
d Explain the importance of fertilisation in sexual reproduction. (16)

4

Biodiversity and natural resources

STARTING POINTS

- There are vast numbers of living things – an almost unlimited diversity. Within this array of organisms are two types of cellular organisation:
  - the prokaryotes (for example, the bacteria), with very small cells without a true nucleus
  - the eukaryotes (for example, plants, animals and fungi), with a nucleus separated from the cytoplasm by a nuclear membrane.
- The fine structure of cells, known as cell ultrastructure, is investigated by the technique of electron microscopy. The animal cell can be viewed as a three-dimensional sac of different organelles, suspended in an aqueous medium – the cytosol – and contained within a plasma membrane.
- Vast numbers of organic compounds make up living things, but most fall into one of four groups of compounds, each with distinctive structures and properties. They are the carbohydrates, lipids, proteins and nucleic acids.
- Histology is the study of the microscopic structure of tissues and organs of multicellular organisms. A tissue is a group of cells of similar structure that perform a particular function. An organ is a collection of tissues and has a specific role.

4.1 The plant cell and its components

1 a Explain why living cells cannot be observed directly in an electron microscope.

  b What basic steps are typically undertaken to produce TEMs of cells?

The fine structure (**ultrastructure**) of living cells was discovered when the electron microscope was applied to biological investigations. You will remember that, while living cells can be examined using the light microscope, this is not the case with the electron microscope. However, there are various techniques by which living tissues can be prepared for observation and analysis by transmission electron microscopy, from which the resulting images can be interpreted with confidence. The results of these observations on animal cells were introduced in Chapter 3, page 100. Now we need to consider the plant cell in the same way, and to compare its structure with that of the animal cell.

Remind yourself of the details of animal cell structure now (Figures 3.3 to 3.10).

The ultrastructure of the plant cell

Images of the ultrastructure of plant cells are based on the interpretations of transmission electron micrographs (TEMs) in just the same way as for animal cells. In Figure 4.1 the ultrastructure of a plant and animal cell is compared by means of a diagram that is half plant cell and half animal cell. This representation of the internal structures makes it clear that these cells have most, but not quite all, of their organelles in common.

The animal cell does have an organelle that is absent from plant cells, namely the **centrosome** (Figure 3.10, page 105). Similarly, a plant cell may contain **chloroplasts**, organelles never found in animal cells. Additionally, all but the very youngest plant cells contain at least one large permanent **vacuole**. These are typically so large they push the cytoplasm with its nucleus and organelles to the cell perimeter (Figure 4.2). Note, too, that the vacuole has a distinctive membrane separating vacuole contents from cytoplasm, called the **tonoplast**.

Finally, an additional feature of every plant cell is the distinctive **cell wall** that surrounds it. These differences are examined in more detail, next. Table 4.1 is a summary of the distinctive features of animal and plant cells.

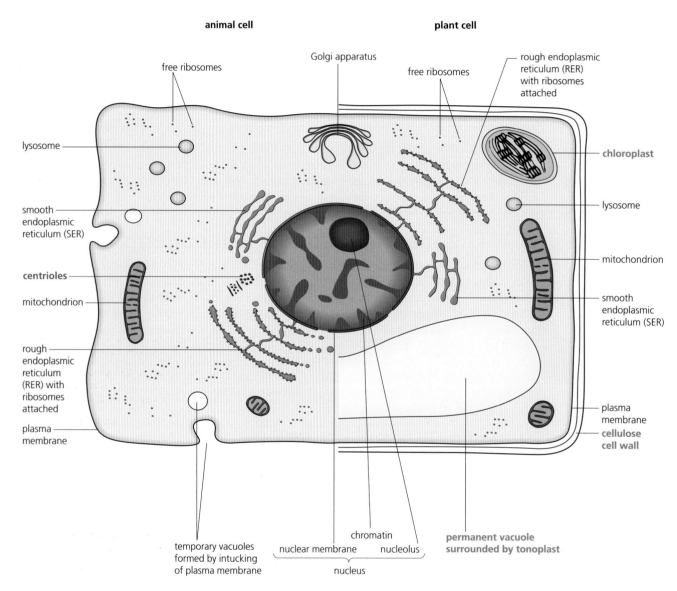

animal cell

plant cell

free ribosomes

Golgi apparatus

free ribosomes

rough endoplasmic reticulum (RER) with ribosomes attached

lysosome

chloroplast

smooth endoplasmic reticulum (SER)

lysosome

centrioles

mitochondrion

mitochondrion

smooth endoplasmic reticulum (SER)

rough endoplasmic reticulum (RER) with ribosomes attached

plasma membrane

cellulose cell wall

plasma membrane

temporary vacuoles formed by intucking of plasma membrane

nuclear membrane

chromatin

nucleolus

permanent vacuole surrounded by tonoplast

nucleus

Figure 4.1 The ultrastructure of the plant and animal cells compared. The labels in green identify structures that occur in plant cells only, while red labels show structures that are unique to animal cells.

Figure 4.2 TEM of a mesophyll cell from a green leaf.

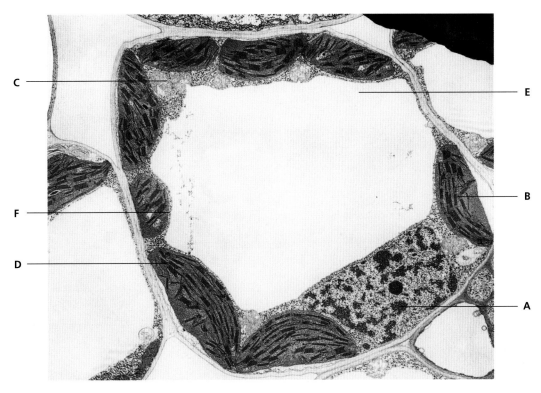

Table 4.1 Key features of animal and plant cells.

Plant cells		Animal cells
	Similarities	
contain cytoplasm surrounded by plasma membrane	**cytoplasm plasma membrane**	contain cytoplasm surrounded by plasma membrane
contain nucleus surrounded by nuclear membrane and containing chromatin and nucleolus	**nucleus**	contain nucleus surrounded by nuclear membrane and containing chromatin and nucleolus
cytoplasm contains all these organelles	**mitochondria endoplasmic reticulum (smooth and rough) Golgi apparatus ribosomes lysosomes**	cytoplasm contains all these organelles
	Differences	
no centrosome present	**centrosome**	a centrosome of two centrioles present outside the nucleus
large, fluid-filled vacuole normally present	**vacuole**	no large permanent vacuoles
cells may contain chloroplasts – the site of photosynthesis (the basis of the autotrophic nutrition of green plants)	**chloroplasts**	no chloroplasts present – animal cells unable to carry out photosynthesis
carbohydrate typically stored as starch, present in organelles called amyloplasts	**carbohydrate storage**	carbohydrate stored as glycogen, often present
cell wall present, largely composed of cellulose, and laid down by the cell around the outside of the plasma membrane	**cell wall**	no cell wall – the plasma membrane forms the outer surface of the cell

2 Identify the organelles labelled A, B, C, D, E and F in Figure 4.2, and state the chief role of each in the living cell.

Chloroplasts and amyloplasts

Chloroplasts are members of a larger group of organelles called **plastids**, found in many plant cells but never in animal cells. Colourless plastids in which starch grains are stored are known as **amyloplasts** (Figure 4.3). Remember, starch is a polysaccharide of two compounds, amylose and amylopectin (Figure 1.6, page 7).

Figure 4.3 Starch storage in amyloplasts.

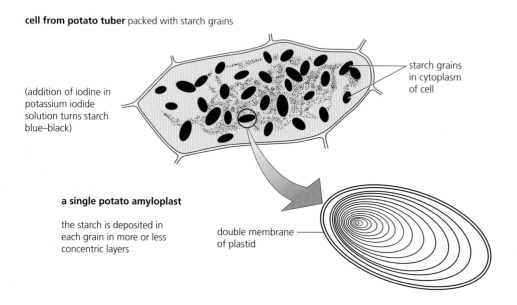

cell from potato tuber packed with starch grains

(addition of iodine in potassium iodide solution turns starch blue–black)

starch grains in cytoplasm of cell

a single potato amyloplast

the starch is deposited in each grain in more or less concentric layers

double membrane of plastid

starch grains take up characteristic shapes in different species

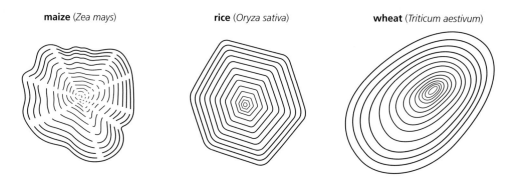

maize (*Zea mays*)   **rice** (*Oryza sativa*)   **wheat** (*Triticum aestivum*)

Activity 4.1: The distribution of starch in plants

Chloroplasts are large plastids, green in colour and typically biconvex in shape. They are about 4–10 µm long and 2–3 µm wide. In green plants, most chloroplasts occur in the mesophyll cells of leaves, but they are also found in the cells of the outer parts of non-woody (herbaceous) stems. A mesophyll cell may be packed with 50 or more chloroplasts. Photosynthesis occurs in chloroplasts, especially in those found in the leaves of plants.

The chloroplast, like all plastids, has a double membrane. In this organelle, the outer layer of the membrane is a continuous boundary, but the inner layer becomes in-tucked to form a system of branching membranes called lamellae or **thylakoids**. In the interior of the chloroplast, the thylakoids are arranged in flattened circular piles called **grana** (singular granum), which look a little like a stack of coins. It is here that the photosynthetic pigments (chlorophylls) are located. There are a large number of grana present, but between them the branching membranes are very loosely arranged in an aqueous matrix, possibly containing small starch grains. This part of the chloroplast is called the **stroma**.

TEM of a thin section of a chloroplast (×15 000)

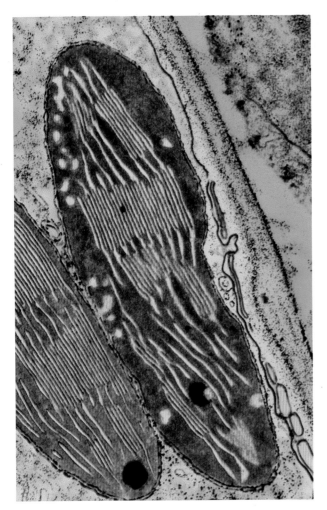

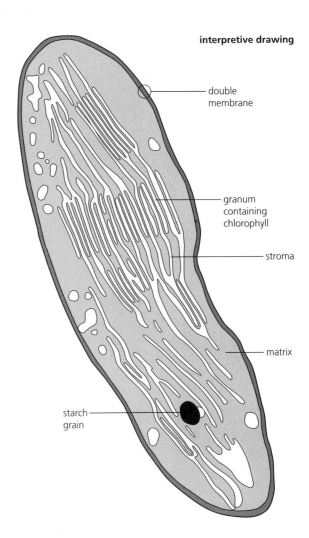

interpretive drawing

double membrane

granum containing chlorophyll

stroma

matrix

starch grain

Figure 4.4 The structure of the chloroplast.

A large, permanent vacuole

Vacuoles are fluid-filled cavities bounded by a single membrane, called the tonoplast. Young plant cells usually contain several small permanent vacuoles which, in the mature cell, have united to form a large, central vacuole, occupying about 80% of the cell volume. The plant cell is filled with a fluid known as **cell sap**, an aqueous solution of dissolved food materials, ions, waste products and sometimes water-soluble pigments. So the vacuole may be considered to function as both a 'larder' and 'dustbin'.

The plant cell wall

The presence of a wall is a characteristic of plant cells. The wall is entirely external to the cell, completely surrounding the plasma membrane. Although not an organelle, the wall is produced by the actions of organelles, as we shall see. Plant cell walls are primarily constructed from cellulose, an extremely strong material.

The chemistry of cellulose

The composition and structure of starch, a polysaccharide, was introduced on page 6. Cellulose, too, is a polysaccharide, also made of glucose molecules combined together. There the similarities end!

Cellulose is a straight-chain (unbranched) polymer of around 2000–3000 glucose units, in which carbon atom 1 of one β-glucose links to carbon atom 4 of the next. Links like this are known as **β-1,4-glycosidic linkages**. This construction results in a long, straight-chain molecule that naturally packs closely together in bundles of about 200 chains, held together by hydrogen bonds, and known as **microfibrils**. The unique strength of cellulose comes from the combined effect of the bonds between the glucose monomers and the **hydrogen bonds** within and between the chains.

Figure 4.5 The chemistry and structure of cellulose.

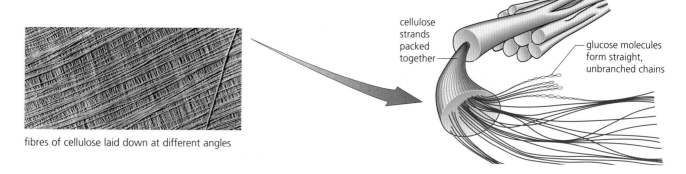

electron micrograph of cellulose in a plant cell wall (×1500)

fibres of cellulose laid down at different angles

cellulose strands packed together

glucose molecules form straight, unbranched chains

3 a What do you understand by *symbiosis*?
b What other examples of symbiosis are you already familiar with?
4 Construct a table of the similarities and differences of starch and cellulose.

The physical properties of cellulose are remarkable. In addition to its extraordinary tensile strength, it is insoluble, tough, durable and slightly elastic. It can be chemically hydrolysed to glucose with difficulty. Certain fungi, bacteria, and possibly some protozoa produce an enzyme known as cellulase that catalyses the hydrolysis of cellulose, forming glucose or sugar acids. Herbivorous animals do not produce cellulase, but harbour in their gut a vast population of symbiotic bacteria that do.

An outcome of these distinctive properties is that cellulose is by far the most abundant carbohydrate – it makes up more than 50% of all organic carbon. In addition to the cellulose of the cell walls of green plants, vast quantities make up the debris of plants in and on the soil (the remains of plant cells, in fact). Thus cellulose is an abundant raw material in the environment, with numerous, important commercial and industrial applications, too. This issue is discussed later in this chapter.

The formation of the cell wall

When a growing plant cell divides, on completion of mitosis, a cell wall is laid down across the old cell, dividing the contents as part of cytokinesis (page 113). The steps involved in wall formation have been observed in a series of electron micrographs taken of dividing plant cells. In the first step, a gel-like layer of calcium pectate, known as the **middle lamella**, is delivered and then deposited by vesicles cut off from the Golgi apparatus and endoplasmic reticulum. These vesicles coalesce along the midline of the dividing cell. This is the first layer of the wall to be deposited. Some of the endoplasmic reticulum of the parent cell becomes trapped across the middle lamella at various points along the developing wall. This persists and they become cytoplasmic connections between the new cells. These connections are called **plasmodesmata**. You can see how plasmodesmata form in Figure 4.6.

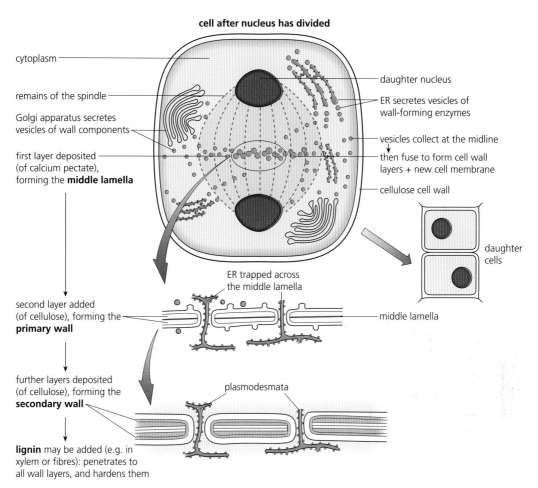

Figure 4.6 The formation of a new plant cell wall.

Onto this middle lamella are then deposited layers of cellulose microfibrils. Initially the microfibrils are typically deposited transversely, as they form what is called the primary cell wall. However, as the cell grows and enlarges, the primary wall microfibrils begin to slide past each other as they re-orientate in response to the strains imposed by the lengthening cell. Something of a meshwork of microfibrils results (Figure 4.7a).

Subsequently, more layers of cellulose are normally added to the inner surface of the wall, forming the secondary cell wall. In secondary walls, cellulose microfibrils are frequently deposited in different layers, often at right angles to each other, forming a meshwork from the outset. The electron micrograph of cellulose microfibrils in Figure 4.5 illustrates this. These additional layers are added after growth in overall size of the cell is completed. The combined effect of all these microfibrils lying at different angles is to greatly increase the strength of the wall.

In some plant cells, the secondary layers of cellulose become very thick indeed – much of the interior of the cell may be taken up by wall material. This is the case in cells called fibres (Figure 4.7b).

In addition, walls may become impregnated with a complex chemical called **lignin**. Lignin hardens and further strengthens the wall, although it is not a carbohydrate. Lignified walls are seen in the water-conducting elements (**xylem** vessels, page 152) and in **fibres** (page 153). Since

5 Explain the difference between a pit and a plasmodesma.

lignin is a hydrophobic material, it makes the bulk of the wall of a xylem vessel more-or-less impervious to water. Consequently the pits that form in xylem vessel walls (and are found in many other plant cells, too) are the main channels by which water may enter and leave xylem vessels during movement through the plant root, stem and leaves. In Figure 4.8 the formation of pits is shown.

a primary wall fibres and their re-orientation as the young cell grows and enlarges

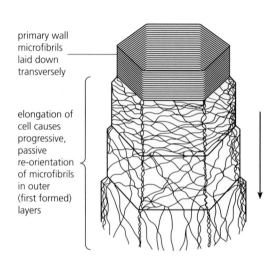

primary wall microfibrils laid down transversely

elongation of cell causes progressive, passive re-orientation of microfibrils in outer (first formed) layers

b primary, secondary and subsequent cell wall layers laid down in a fibre's cell wall

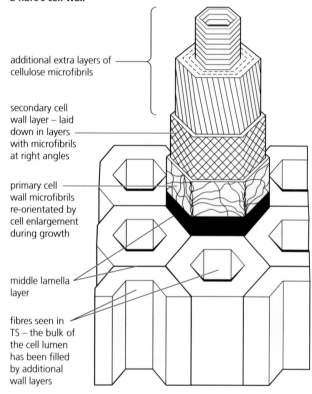

additional extra layers of cellulose microfibrils

secondary cell wall layer – laid down in layers with microfibrils at right angles

primary cell wall microfibrils re-orientated by cell enlargement during growth

middle lamella layer

fibres seen in TS – the bulk of the cell lumen has been filled by additional wall layers

Figure 4.7 Primary and secondary cell wall structure.

Figure 4.8 The formation of pits in the cell wall.

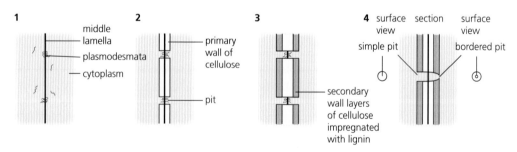

1

middle
lamella

plasmodesmata

cytoplasm

2

primary
wall of
cellulose

pit

3

secondary
wall layers
of cellulose
impregnated
with lignin

4 surface   section   surface
view   view

simple pit   bordered pit

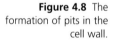

Activity 4.2: Finding out about pectin

Activity 4.3: Ultrastructure of named plant and animal cells compared

■ Extension: Pectins

In addition to cellulose, plant cell walls contain other polysaccharides which contribute to their mechanical strength. The chief examples are pectins, and these are polymers of another sugar (called galactose) and its sugar acid. Pectins become bound together by calcium ions (forming calcium pectate – notably present in the middle lamella layer).

4.2 The working plant

Green plants make up one of the five kingdoms of living things (page 172), but within this kingdom, the flowering plants are the dominant terrestrial plants in almost every habitat across the world today. Some flowering plants are trees and shrubs with woody stems, but many are non-woody (herbaceous) plants. Whether woody or herbaceous, the flowering plant consists of stem, leaves and root.

- The stem supports the leaves in the sunlight, and transports organic materials (for example, sugar, amino acids), ions and water between the roots and leaves. At the top of the stem and in the axil of each leaf are growing points protected in buds. New cells are produced at these growing points at certain times of the year.
- A leaf consists of a leaf blade connected to the stem by a leaf stalk. The leaf is an organ specialised for photosynthesis.
- The root anchors the plant and is the site of absorption of water and ions from the soil.

The structure of the sunflower, *Helianthus annuus*, a herbaceous plant, illustrates these features (Figure 4.9).

Figure 4.9 The structure of the sunflower plant, *Helianthus annuus*.

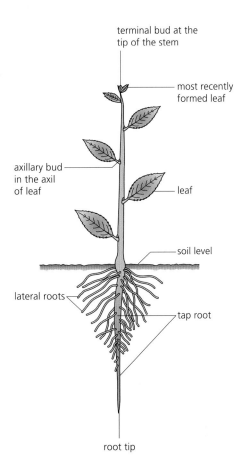

A* Extension 4.1: Exploring the plant kingdom

The distribution of tissues in the sunflower

We can best show the distribution of tissues in a plant organ by means of a **tissue map**. A tissue map (sometimes called a 'low power diagram') is a drawing that records the relative positions of component parts within an organ, as seen in section. The tissue map does not show individual cells. Figure 4.10 shows examples of tissue map drawings (see also Activity 4.4, page 154).

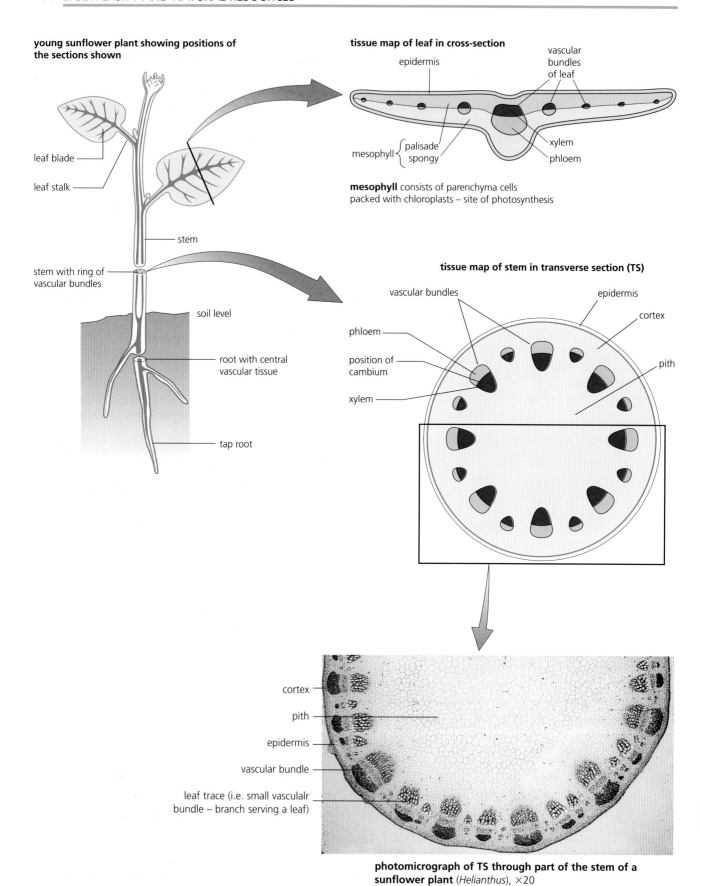

young sunflower plant showing positions of the sections shown

leaf blade

leaf stalk

stem

stem with ring of vascular bundles

soil level

root with central vascular tissue

tap root

tissue map of leaf in cross-section

epidermis

vascular bundles of leaf

mesophyll { palisade / spongy }

xylem

phloem

mesophyll consists of parenchyma cells packed with chloroplasts – site of photosynthesis

tissue map of stem in transverse section (TS)

vascular bundles

epidermis

cortex

phloem

position of cambium

xylem

pith

cortex

pith

epidermis

vascular bundle

leaf trace (i.e. small vasculalr bundle – branch serving a leaf)

photomicrograph of TS through part of the stem of a sunflower plant (*Helianthus*), ×20

Figure 4.10 The distribution of tissues in the plant.

Stem

From the tissue maps in Figure 4.10, it can be seen that the stem is an organ surrounded or contained by a single layer of cells called the **epidermis**, and that it contains **vascular tissue** (**xylem** and **phloem** tissues) in a discrete system of veins or vascular bundles. In the stem, the vascular bundles are arranged in a ring, positioned towards the outside of the stem, rather like the steel girders of a ferro-concrete building. Vascular bundles clearly have a support role in the stem and leaves (see below).

Another role of stem tissues is internal transport. Water is transported in the xylem from the roots to the whole aerial system (stems and leaves). Dissolved inorganic ions are carried up in the water in the xylem, too. Xylem tissue is, in effect, a system of hollow tubes. Organic solutes like sugars and amino acids are distributed in the phloem to growing points throughout the plant. Within phloem tissue as a whole, transport occurs both up and down the plant. Phloem tissues consist of intact cells with living contents, in complete contrast to xylem vessels.

Leaf

A tissue map showing the distribution of tissues in a leaf is shown in Figure 4.10. As with the stem, the leaf is contained by a single layer of cells, the epidermis, and also contains vascular tissue in a system of vascular bundles. The vascular bundles in leaves are often referred to as veins. The bulk of the leaf is taken up by a tissue called **mesophyll**, and the cells here are supported by veins arranged in a branching network.

The leaf is an organ specialised for photosynthesis.

Figure 4.11 Turgidity supports plants.

a plant cell with adequate supply of water

in this state, all cell walls exert pressure on the surrounding cell walls and the tissue is turgid – fully supported

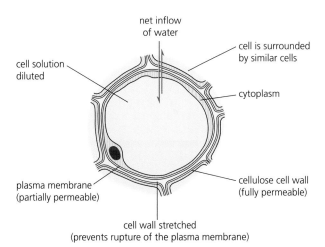

net inflow of water

cell solution diluted

cell is surrounded by similar cells

cytoplasm

plasma membrane (partially permeable)

cellulose cell wall (fully permeable)

cell wall stretched (prevents rupture of the plasma membrane)

b wilting

well-watered potted plant – all cells are turgid

plant exposed to drought conditions, to the point of wilting – water loss by evaporation has turned turgid cells flaccid

Support in the plant stem

Support in the herbaceous plant comes firstly from the **turgidity** of all the cells of the packing tissue, known as **parenchyma**. Plant cells are normally turgid, with their vacuoles engorged with dilute cell sap, and with their cell contents pushed hard against the walls. The walls, which hardly stretch, in turn push hard against each other (turgor pressure, Figure 4.11a), thereby generating rigidity.

Of course, all the packing cells are tightly contained within the tough epidermis. The epidermis, therefore, contributes to mechanical support, too (as well as giving some protection against entry of parasites such as fungi, for example).

Immediately below the epidermis a different form of packing tissue, known as **collenchyma** tissue, is often found. These cells, otherwise very similar to parenchyma, have additional layers of cellulose at the corners of the cells (Figure 4.13). Collenchyma consists of flexible supporting cells, and typically occurs in leaves as well as stems.

The importance of the turgidity of all these living cells in the support of stem and leaves is demonstrated whenever a herbaceous plant is unfortunately deprived of water to the point that it wilts (Figure 4.11b).

The stem is also supported by the ring of vascular bundles and leaf traces (these supply the leaves attached to the stem at intervals). This vascular tissue contains two types of cells with lignified walls.

- **Xylem vessels** (Figure 4.12) are hollow tubes, with lateral walls reinforced with additional layers of cellulose, impregnated with lignin during development. Xylem vessels also have pit areas in their walls which allow the lateral movement of water to surrounding cells and tissues.
- **Fibres (sclerenchyma)** (Figure 4.13) form an outer cap to the vascular bundle in the sunflower stem. By the time fibres have finished growing, they are dead, empty cells with tapering, interlocking walls greatly thickened by additional layers of cellulose, impregnated with lignin as the fibres formed. Fibres are also found between xylem vessels, thereby strengthening the whole bundle.

Figure 4.12 The structure of xylem tissue.

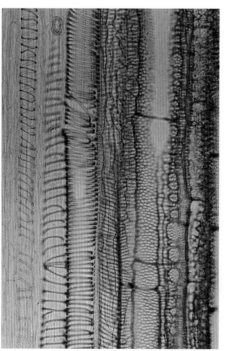

photomicrograph of xylem tissue in LS

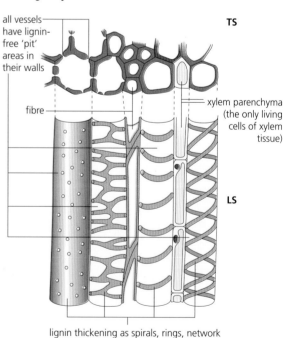

drawing of xylem vessels in TS and LS

all vessels have lignin-free 'pit' areas in their walls

TS

fibre

xylem parenchyma (the only living cells of xylem tissue)

LS

lignin thickening as spirals, rings, network or solid blocks – deposited on inside of vessel, strengthening the cellulose layers

6 By means of a table, compare the similarities in and differences between the positions, structures and functions of xylem and sclerenchyma in the herbaceous plant stem.

The plant stem can be likened to a modern high-rise ferro-concrete building, where the hard, inextensible, interconnecting steel girders (vascular bundles, Figure 4.14) are surrounded by the softer, incompressible concrete (parenchyma). Of course, the stem has significant flexibility, allowing bending in the wind, so lessening the chance of damage.

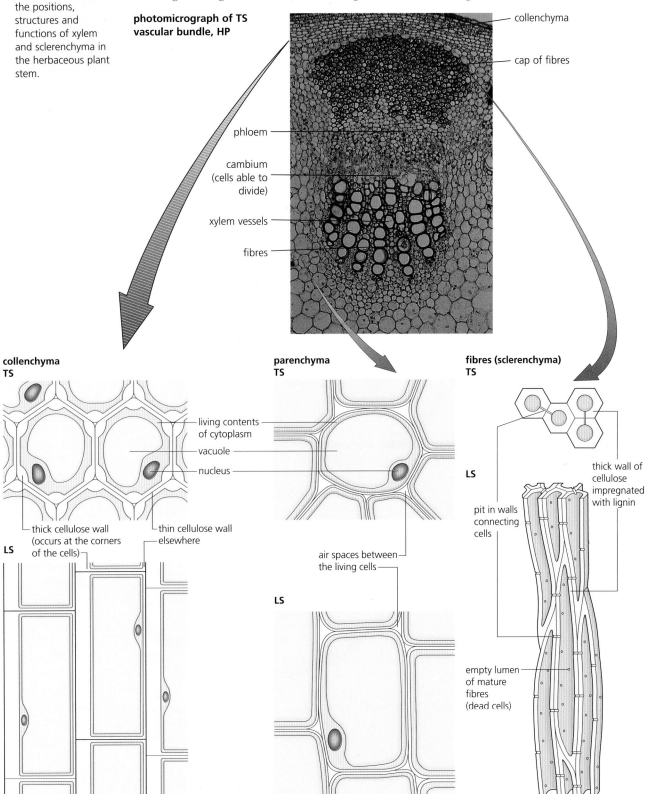

photomicrograph of TS vascular bundle, HP

collenchyma

cap of fibres

phloem

cambium (cells able to divide)

xylem vessels

fibres

collenchyma TS

parenchyma TS

fibres (sclerenchyma) TS

living contents of cytoplasm

vacuole

nucleus

thick cellulose wall (occurs at the corners of the cells)

thin cellulose wall elsewhere

LS

air spaces between the living cells

LS

LS

pit in walls connecting cells

thick wall of cellulose impregnated with lignin

empty lumen of mature fibres (dead cells)

Figure 4.13 Support tissues around a bundle.

the leaf traces branch from the vascular bundles, to the leaves

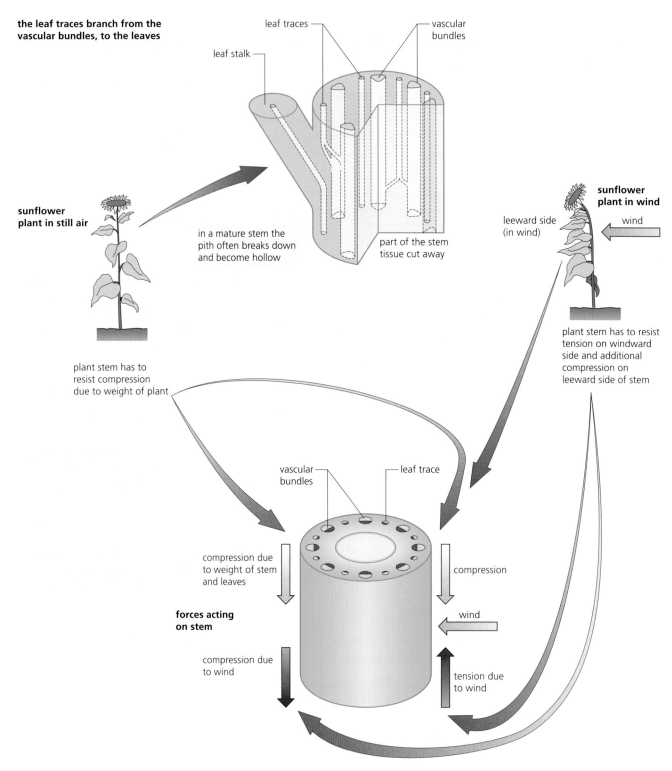

Figure 4.14 Forces the stem resists.

Activity 4.4: Creating tissue maps and HP drawings

Figure 4.15 Annual meadow grass (*Poa* sp.) shows monocotyledonous features.

Activity 4.5: Monocotyledons and dicotyledons – the differences

A* Extension 4.2: The histology of monocotyledonous plants

■ Extension: Dicotyledonous and monocotyledonous plants

The flowering plants are divided to two groups, known as the monocotyledons and dicotyledons, with distinguishing structural features. The sunflower is a typical dicotyledonous plant (Figure 4.9) and the grasses are a major family of monocotyledonous plants (Figure 4.15). Monocotyledons have bayonet- or strap-shaped leaves with parallel veins. Actually, the defining difference between members of the two groups is the number of seed leaves (cotyledons) present in the embryo. (Cotyledons are structurally simpler leaves than the normal leaves of these plants that develop later.)

■ Investigation: Determining the tensile strength of fibres from nettle stems

Flax (*Linum usitatissimum*) and certain other plant species have been cultivated for more than 4000 years in order to obtain fibres. Flax fibres are especially strong yet pliable, and are used in the production of linen for making into garments. The harvested plants are typically laid under water to soften the stem tissues and release the fibres by the action of micro-organisms – a process called 'retting'. These processes can be simulated in the laboratory retting of plant fibres from common plants such as nettles, as in the investigation outlined below.

The investigation requires the:

1 extraction of bundles of fibres from the stems of plants such as the common nettle (*Urtica dioica*), and the preparation of the fibres for measurement of their tensile strength
2 design of an appropriate laboratory bench test to estimate the tensile strength of these bundles (for example, Figure 4.16)
3 detailed consideration of safety issues that will arise, and the essential steps required to avoid unnecessary risks when this investigation is conducted as a group project, for example.

Step 1 involves soaking stems freed from leaves and flowers in a bucket of water for at least a week. Softened by this treatment, the other stem tissues around the bundles may be rubbed away manually, and the bundles of fibres prepared by drying. It may be appropriate for this stage to have been completed beforehand, by laboratory staff.

Step 2 requires determination of the maximum stress (pulling force) the fibres can withstand. Tensile strength is expressed in $N\,m^{-2}$, so in addition to recording the force applied, and the effects this has on the fibres, the typical cross-sectional area of the nettle bundles of fibres must be estimated.

Figure 4.16 A simple 'tensometer' apparatus and the likely profile of a plot of force against extension in fibre length.

HSW 4.1: Criteria 2, 3, 4, 5 and 8 – Extraction of fibres from nettle stems, and measurement of their tensile strength

simple 'tensometer' apparatus

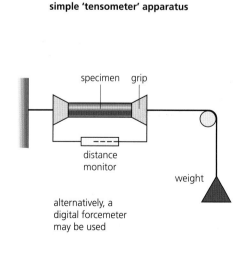

profile of the likely force v. extension plot

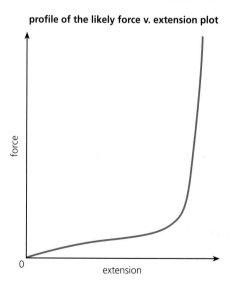

The importance of water to plants

Plants require adequate water because:

- it makes up about 80% of a plant cell
- it is the environment for all the chemical reactions of life
- the turgidity of living cells of the plant plays an important part in support
- the pores in leaves (stomata), which permit the diffusion of carbon dioxide required for photosynthesis, are open only when they are turgid
- water readily evaporates from moist surfaces within the plant under a wide range of external conditions; it must be constantly replaced if wilting is to be avoided.

For these reasons water is essential for plant life.

Water loss and water uptake

The external surface of the stem and leaves are covered by a layer of wax called the **cuticle** which effectively prevents water vapour loss by evaporation from the cells of the outer surfaces of the aerial system, even in dry, warm and windy weather.

Figure 4.17 Water loss from the aerial system of the plant.

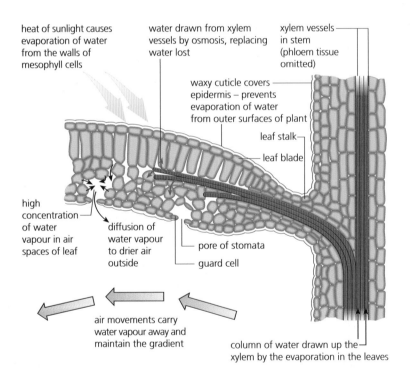

heat of sunlight causes evaporation of water from the walls of mesophyll cells

water drawn from xylem vessels by osmosis, replacing water lost

xylem vessels in stem (phloem tissue omitted)

waxy cuticle covers epidermis – prevents evaporation of water from outer surfaces of plant

leaf stalk

leaf blade

high concentration of water vapour in air spaces of leaf

diffusion of water vapour to drier air outside

pore of stomata

guard cell

air movements carry water vapour away and maintain the gradient

column of water drawn up the xylem by the evaporation in the leaves

7 Explain precisely why dry, warm and windy weather conditions favour evaporation of water.

However, water does evaporate from the cell walls of all the mesophyll cells of the leaf, and the result is that water vapour accumulates in the air spaces of the plant. These spaces are continuous, and they also connect with the chambers below the stomatal pores that are present in large numbers in the epidermis of leaves.

Stomata are normally open in the light, thereby facilitating photosynthesis. Water vapour inevitably diffuses out through the open stomata. In the light, the absorption of which is essential for photosynthesis, the temperature of cells is raised, and this speeds up evaporation further. Incidentally, evaporation of water cools the leaf cells.

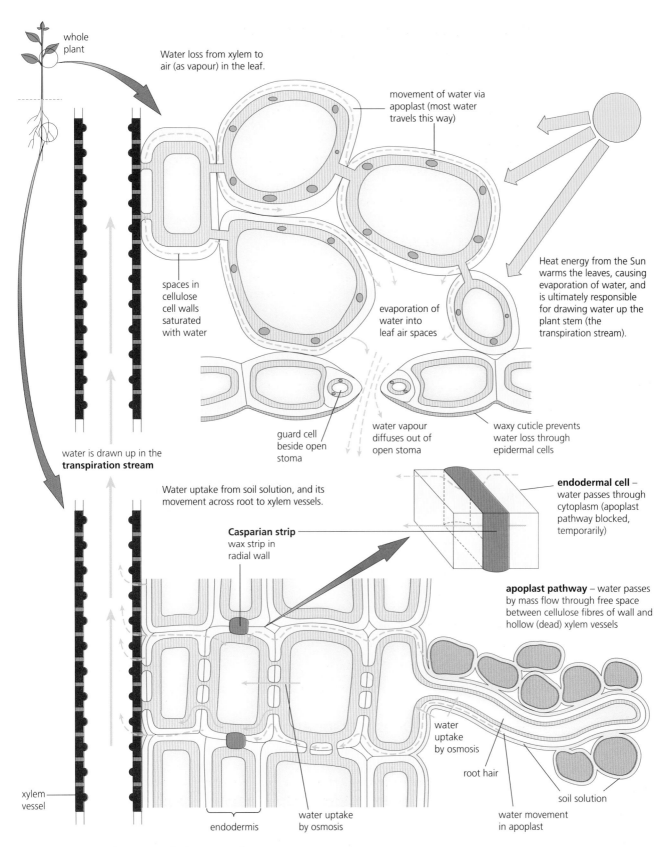

whole plant

Water loss from xylem to air (as vapour) in the leaf.

movement of water via apoplast (most water travels this way)

spaces in cellulose cell walls saturated with water

evaporation of water into leaf air spaces

Heat energy from the Sun warms the leaves, causing evaporation of water, and is ultimately responsible for drawing water up the plant stem (the transpiration stream).

guard cell beside open stoma

water vapour diffuses out of open stoma

waxy cuticle prevents water loss through epidermal cells

water is drawn up in the **transpiration stream**

Water uptake from soil solution, and its movement across root to xylem vessels.

endodermal cell – water passes through cytoplasm (apoplast pathway blocked, temporarily)

Casparian strip wax strip in radial wall

apoplast pathway – water passes by mass flow through free space between cellulose fibres of wall and hollow (dead) xylem vessels

water uptake by osmosis

root hair

soil solution

xylem vessel

endodermis

water uptake by osmosis

water movement in apoplast

Figure 4.18 Water loss and uptake by a green plant – a summary.

Transpiration is the name given to the process by which water is lost as vapour from the aerial parts of plants (Figure 4.17).

As a result of transpiration, water is drawn from the xylem vessels of the leaf veins to replace the lost water from the leaf cells. Consequently, a stream of water is drawn up through the plant from the roots; many litres are moved in this way on a hot day. Even in tall trees, columns of water are drawn up in what is referred to as the transpiration stream. The water column does not break or pull apart, for liquid water has remarkable adherence and tension resistance properties ('cohesion–tension theory'). These qualities of water are attributable to the hydrogen bonds formed between water molecules (page 2).

The mechanisms of water absorption from the soil solution, the evaporation of water from leaves in the light, and the resulting transpiration stream are summarised in Figure 4.18. When a water column is under tension in a tube, the tendency is for the walls of the tube to collapse. Consequently, we would expect the thickenings of the walls of xylem vessels to be laid down inside.

Examine Figure 4.19 and decide if this is so.

Figure 4.19 SEM of xylem vessels.

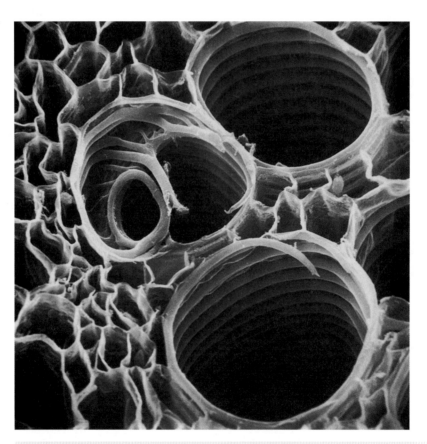

■ Extension: Does transpiration have a role?

Transpiration is a direct consequence of plant structure, plant nutrition, and the mechanism of gas exchange in leaves. In effect, the living plant is a 'wick' that steadily dries the soil around it. Put like this, transpiration is an unfortunate consequence of plant structure and metabolism, rather than a valuable process!

However, an effect of evaporation is that the leaf cells are cooled in the light. Furthermore, the stream of water travelling up from the roots passively carries the dissolved ions that have been absorbed from the soil solution in the root hairs and are required in the leaves and growing points of the plant. So, transpiration has a role in cooling the plant in sunlight, and also in moving mineral ions up the stem. This latter point is our next issue to discuss.

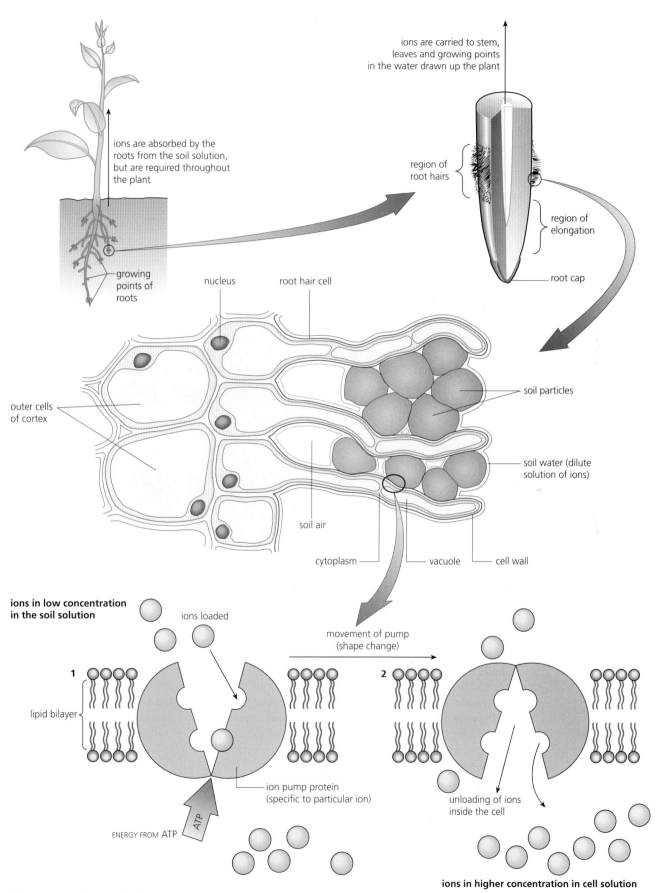

ions are carried to stem,
leaves and growing points
in the water drawn up the plant

ions are absorbed by the
roots from the soil solution,
but are required throughout
the plant

region of
root hairs

region of
elongation

root cap

growing
points of
roots

nucleus

root hair cell

soil particles

outer cells
of cortex

soil water (dilute
solution of ions)

soil air

cytoplasm

vacuole

cell wall

**ions in low concentration
in the soil solution**

ions loaded

movement of pump
(shape change)

1

2

lipid bilayer

ion pump protein
(specific to particular ion)

unloading of ions
inside the cell

ENERGY FROM ATP

ATP

ions in higher concentration in cell solution

Figure 4.20 Active uptake of ions in roots.

The importance of inorganic ions to plants

Living things need a wide range of inorganic ions, but only a handful are needed in relatively large quantities. These are known as the major mineral elements, or **macronutrients**. Major mineral elements include cations like calcium (Ca^{2+}), iron (Fe^{2+} and Fe^{3+}), and magnesium (Mg^{2+}), and anions, like phosphate ($H_2PO_4^-$), nitrate (NO^{3-}) and sulphate (SO_4^{2-}).

The roles of major mineral elements may be as structural components of essential substances, including calcium for plant cell walls, phosphates in nucleic acid and ATP, nitrogen and sulphur in amino acids, peptides and proteins, and magnesium in chlorophyll. Major elements also play a part in the activities of cells, including water uptake by osmosis.

Other inorganic elements, required in tiny amounts, are referred to as trace elements or **micronutrients**. Examples include metal ions like copper (Cu^{2+}) and zinc (Zn^{2+}) and the non-metal ions like iodine (I^-) and molybdenum (Mo_4^{2-}). The roles of micronutrients may be as co-factors in the catalytic activity of particular enzymes.

Absorption of mineral ions for plant metabolism

The source of mineral ions for plants is the soil solution. This is a layer of water that occurs around the individual soil particles. Here, very many different ions occur, most in quite low concentrations. Plants can take up ions from the soil solution through their roots. There is a region of the root, near the root tip, where most if not all of the absorption from the soil solution occurs. This is called the region of root hairs.

Root hairs form just behind the region where growth in length of the root takes place. These hairs are extremely delicate structures, tiny extensions of the outermost layer of cells. If they were formed where the root was still pushing its way between the soil particles, the root hairs would be quickly torn off. Further back, the root becomes impervious, and the outer layers of living cells (root hairs included) are sloughed off.

Water uptake from the soil and the absorption of mineral ions occurs at the root hairs, but by entirely different mechanisms. Water uptake is discussed above. Ions are taken up selectively, by an active process, using energy from respiration (page 51). This means that, while many of the available ions are accumulated, others largely remain outside the cells of the root. The uptake process occurs by means of protein pumps in the plasma membranes of the root cells, including root hair cells. Protein pumps are specific for particular ions; only if a particular transport protein is produced by the cells and built into the cell membranes can uptake occur. Then if the ion is available it may be actively accumulated, whether or not there is a higher concentration in the cell already.

8 Why and how is the supply of essential ions maintained in soil in horticulture?

■ Investigation: Investigating mineral deficiency practically

The growing of terrestrial plants with roots in an aerated solution is known as water culture or hydroponics. This technique can be used to discover which ions are essential for normal healthy plant growth (Figure 4.21).

A 'balanced' solution (complete culture solution) is one that provides all the necessary ions at appropriate concentrations. Solutions for water culture experiments are prepared from extremely pure chemicals dissolved in deionised water. Carefully cleaned glassware is essential. Many grass species (including a cultivated grass such as barley) are suitable for studies of ion deficiency, for these plants depend almost exclusively on the external supply, because their seeds are small and contain only limited reserves of ions.

The effects of deficiencies of specific ions can be studied using this water culture method. Alternative culture solutions are prepared, each having one selected element absent. Plants grown in deficient media are compared with control plants grown in complete culture solution. Incidentally, knowledge of the deficiency symptoms commonly shown by plants lacking an adequate supply of an essential ion is important in horticulture and agriculture.

This water culture technique can be adapted to use duckweed (*Lemna* sp.), floating on culture solution in a Petri dish.

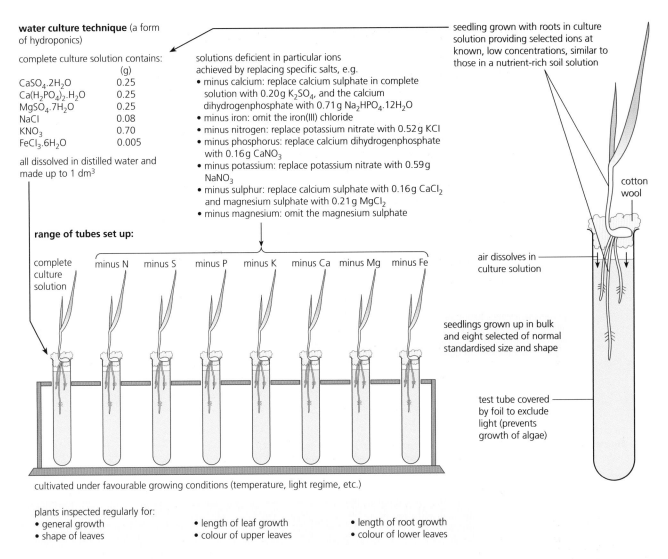

water culture technique (a form of hydroponics)

complete culture solution contains:

	(g)
$CaSO_4.2H_2O$	0.25
$Ca(H_2PO_4)_2.H_2O$	0.25
$MgSO_4.7H_2O$	0.25
NaCl	0.08
KNO_3	0.70
$FeCl_3.6H_2O$	0.005

all dissolved in distilled water and made up to 1 dm³

solutions deficient in particular ions achieved by replacing specific salts, e.g.
- minus calcium: replace calcium sulphate in complete solution with 0.20 g K_2SO_4, and the calcium dihydrogenphosphate with 0.71 g $Na_2HPO_4.12H_2O$
- minus iron: omit the iron(III) chloride
- minus nitrogen: replace potassium nitrate with 0.52 g KCl
- minus phosphorus: replace calcium dihydrogenphosphate with 0.16 g $CaNO_3$
- minus potassium: replace potassium nitrate with 0.59 g $NaNO_3$
- minus sulphur: replace calcium sulphate with 0.16 g $CaCl_2$ and magnesium sulphate with 0.21 g $MgCl_2$
- minus magnesium: omit the magnesium sulphate

seedling grown with roots in culture solution providing selected ions at known, low concentrations, similar to those in a nutrient-rich soil solution

cotton wool

range of tubes set up:

complete culture solution | minus N | minus S | minus P | minus K | minus Ca | minus Mg | minus Fe

air dissolves in culture solution

seedlings grown up in bulk and eight selected of normal standardised size and shape

test tube covered by foil to exclude light (prevents growth of algae)

cultivated under favourable growing conditions (temperature, light regime, etc.)

plants inspected regularly for:
- general growth
- shape of leaves
- length of leaf growth
- colour of upper leaves
- length of root growth
- colour of lower leaves

Figure 4.21 Investigating essential ions for plant metabolism.

Experiment: To detect the effects of specific mineral ion deficiencies on plant growth.

Risk assessment: Good laboratory practice is sufficient to avoid a hazard.

Resources:
To carry out the investigation, each student group requires:

- a supply of 50 *Lemna* plants of the same species, in pond water
- nine Petri dishes, 20 cm³ graduated pipette with safety bulb, supply of distilled water
- waterproof marker pen, fine paint brush, and hand lens
- access to complete culture solution (Sacchs formula), and to labelled bottles of culture solutions with specific ions missing, made from the complete culture solution formulation, but with specific substitutions to exclude one of the ions in turn (Figure 4.21).

Collection of suitable *Lemna* plants:
Lemna is a genus of simple aquatic plants consisting of one or more tiny pendulous roots and small green leaves. *Lemna* may grow quickly, forming a green carpet over the surface of stagnant water. Dipping into the surface layers of such a pond with a suitable net yields plenty of these

plants to bring back to the laboratory in pond water. This experiment should be carried out with *Lemna* plants all of the same species. Each student group will require about 50 plants. Figure 4.22 is designed to aid identification.

Method:
Take nine Petri dishes and label them:

- complete culture solution
- minus phosphorus
- minus magnesium
- minus nitrogen
- minus potassium
- minus iron
- minus sulphur
- minus calcium
- distilled water.

Into each dish, measure out 15 cm³ of the appropriate culture solution (or water) provided. Then count out five healthy *Lemna* plants of comparable size into each dish. If any plant has a small 'bud' developing, this should be trimmed away with fine scissors. Replace the lids of the dishes and incubate them all together, in a position that receives good, even illumination.

The experiment should be allowed to proceed for several weeks. If there is a tendency for the medium to evaporate after some weeks, restore the liquid level using distilled water.

Examine the plants about twice each week, using a hand lens. The plants can best be handled by means of a paint brush. For each dish record the following numerical information:

- the number of live plants
- the number of green leaves
- the number of dead leaves (those turning yellow, or white)
- the length of the longest root, recorded either weekly or on alternate weeks, depending on growth rate.

Figure 4.22 Key to species of *Lemna*.

1 leaves without stalks; plants float on surface of water	**2**
leaves develop stalks; plants float below the surface	*L. trisulca*
2 each plant has a single root only	**3**
each plant has several roots	*L. polyrrhiza*
3 plant leaves are thin and flat	*L. minor*
plant leaves are convex below	*L. gibba*

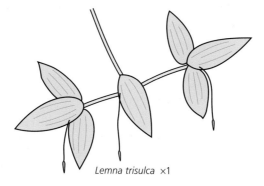

Lemna trisulca ×1

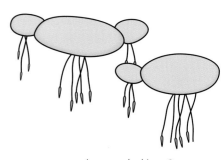

Lemna polyrrhiza ×2

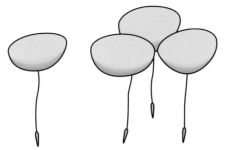

Lemna gibba ×2

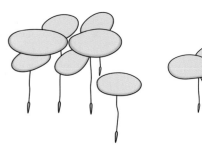

Lemna minor ×3

HSW 4.2: Criteria 2, 3, 4, 5 and 8 – The effects of specific mineral ion deficiencies on plant growth

In addition, look for and record evidence of abnormal growth and development in the absence of any of the mineral nutrient ions.

Data presentation: The bulk of the data generated are numerical, and can be summarised by means of a column graph. Make a listing of any growth abnormalities that appear to be associated with particular deficiencies.

Conclusion: As far as possible, relate the evidence obtained to the published accounts of the ions involved in plant growth, development and metabolism.

Evaluation: What are the advantages and disadvantages of using *Lemna* as an experimental plant in place of a more 'traditional' terrestrial flowering plant? Which of the two dishes, distilled water or complete culture solution, do you see as the 'control' dish?

4.3 Plants with economically important roles

The Industrial Revolution in the developed countries of the world, which began about 200 years ago, has been sustained by fossil fuels. Coal, and later oil, became the energy sources for industry and transport, and sustained the urban living conditions of the huge labour forces on which industrial production was dependent. Agriculture, too, was revolutionised by the ready availability of these fuels.

'Cheap' oil, as it appeared, also serves increasingly as a raw material for a range of products, many of which have replaced products previously derived exclusively from plants.

Fossil fuels were laid down in the long Carboniferous Period, when vast amounts of atmospheric carbon dioxide were locked into organic molecules by photosynthesis in the 'coal' forest plants which then became fossilised. It is this carbon, from the atmospheres of approximately 355 million years ago, that is now being returned to our atmosphere as additional carbon dioxide as fossil fuels are burned. One outcome, namely the enhancement of global warming, makes continuing total dependence on fossil fuel inadvisable. Also, the stocks of these fuels are finite. For these reasons, there is renewed urgency in finding sustainable uses of living plants to meet society's diverse needs for energy and materials.

Figure 4.23 How we use plants.

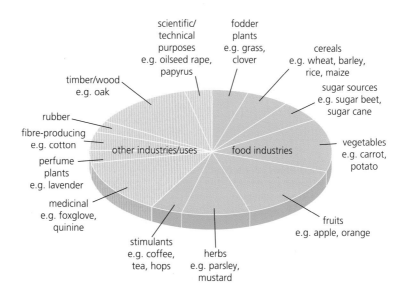

We use plants not only as foods, but also as raw materials for industries such as forestry, medicine, fabric and clothing manufacture, dyeing, and many other industrial, scientific and technical uses. These are summarised in Figure 4.23, where the divisions are comparative, based on the numbers of most commonly exploited species. Incidentally, there is no reference there to the huge numbers of decorative plants used in gardens and parklands, nor to the vast, worldwide uses of plant matter as fuel. We now examine some examples of developments in the economic exploitation of plants.

Case study 1 – Applications of plant fibres

Sclerenchyma (page 152) is the chief form of plant fibre used commercially. This natural product consists chemically of cellulose impregnated with lignin. It has a very long history of exploitation by human communities, but sclerenchyma is not the only form of fibre used. Table 4.2 identifies the types of plant fibres exploited and lists their sources and their important, continuing applications.

In the relatively recent period of cheap and readily available oil reserves, industry has manufactured synthetic polymers similar to cellulose, such as rayon, nylon, acrylics, polyesters and polypropylenes. The 'wrinkle resistant' characteristic of these, when woven into fabrics, has caused them to rival natural fibres like cotton. They continue to dominate in the fibre market, though often in blends made with natural fibres. However, many of the natural fibres listed below have had long-term commercial applications, and their uses are likely to be increasingly sustainable in future.

Table 4.2 Economically important plant fibres and their uses. Note that the term 'fibres' is used in commercial circles for 'fibre-like' structures – *not all are sclerenchyma tissue.*

Category of fibre	Examples	Botanical origins	Uses
surface fibres, from fruits and seeds (non-lignified cells)	cotton (*Gossypium*)	hairs on seeds	textile fibres, to weave cotton
	kapok (*Ceiba*)	hairs within capsules	filling fibres, used for stuffing
bast fibres, from around vascular tissue/under the epidermis of stems/in leaves (sclerencyhma)	flax (*Linum*)	stem	textile fibres to weave cloth
	hemp (*Cannabis sativa*)	stem and leaves	brush fibres, for carpet backing cordage fibres, for marine ropes and twine (string)
	jute (*Cocchorus*)	stem sclerenchyma	brush fibres, to weave mats and sacking
hard fibres, of whole vascular strands from monocotyledonous stems or leaves or fruits (sclerencyhma)	manila hemp (*Musa*) sisal (*Agave*)	leaves	brush fibres, to weave mats and sacking, make baskets and brooms cordage fibres, for marine ropes and twine (string)
	coconut (*Cocos*)	outer wall of coconut fruit	brush fibres, to weave mats and sacking
wood fibres, of shredded or ground-up woods (often from softwood species) (sclerencyhma)	poplar (*Populus*) spruce (*Picea*) and many other softwood and some hardwood tree woods	woody stem	paper making miscellaneous fibres, for mulches in horticulture, cardboards, or compression into chipboards and other construction materials

Activity 4.6: The fibres in your clothes!

9 What are the structural differences between a fibre (sclerenchyma) and a surface hair of plants?

Plant fibres have also become source materials for plastics, and as sources of power or fuel.

■ **Composite plastics may contain plant fibres** from wood, flax and hemp (as replacements for glass fibre) that have been heated with other components, moulded and cooled during their manufacture. They have extensive uses in the production of motor vehicle interior fittings, garden furniture, baseboards, and house-building items such as door frames, for example.

■ **Ethanol fuel or electricity may be generated from agricultural waste fibre** (for example, straw), after pre-treatment to increase surface area, and then hydrolysis using a cellulase of bacterial origin. The resulting sugar solution, fermented by yeast, yields alcohol which may be distilled. Alcohol may be burned to generate electricity, or added to petrol to form a fuel suitable for the internal combustion engine. Schemes of this sort have to be conducted economically, in circumstances where production costs do not exceed the gain from using the product.

Case study 2 – Starch as a resource in place of oil-derived chemicals

Starch is a cheap and abundant natural polymer (page 7), usually extracted from cereal crops and tubers such as the potato. This renewable raw material already has many industrial applications in non-food sectors. For example, starch is used in cardboard, paper and textile manufacture, in building materials, adhesives, cleaning products, cosmetics and pharmaceuticals. In the burgeoning biotechnology industry, it may be used as a source of other nutrients for the micro-organisms being cultured in fermenter tanks. In many of these applications, starch is a preferable alternative to oil-based chemicals that have similar properties, where these are available.

The use of plastics derived from petrochemicals as packaging materials for sale is accelerating worldwide. For example, it is already a major feature of developed countries' marketing, and in countries like India and China that have populations of very many millions, where there are fast-growing urban societies.

The outcome is a vast quantity of plastic packaging waste that has to be disposed of. Put into landfill sites (or allowed to escape into our immediate environment) it may take up to 400 years to be decayed! If burned, it often generates toxic air pollution. It is obviously essential that cheap and easily biodegradable packaging is made available for use in place of petrochemical polymers.

However, starch has limitations as a raw material for this purpose. It is hydrophilic and partially water-soluble, and when untreated starch is heated, it degrades before it can be melted – that is, starch is not thermoplastic.

Starch has been successfully incorporated into more traditional plastics from petrochemical sources to make a significantly more easily biodegraded product – a token improvement. Also, polymers derived from starch have been compounded with other biodegradable substances to form a totally biodegradable plastic. So far, this later development is an expensive alternative, unfortunately.

Research continues to try and find a starch-derived plastic that has appropriate qualities as a secure packaging material for a wide range of dry and 'wet' (that is, food) goods, and yet is both cheap and easily biodegraded by the metabolism of soil micro-organisms, when finally composted.

Case study 3 – Chemical warfare by plants!

Animals may quickly move away from danger; rooted to the spot, plants must remain and are vulnerable! We have already noted that a plant is harmed by severe wilting if the local water supply fails (page 151); plants are also easy targets for browsing herbivores. Not surprisingly, some plants have evolved chemical defence mechanisms. These involve the synthesis and storage in their cells of substances that are harmful when consumed by an invading herbivore or pathogen. The bracken frond (leaf) is an extreme example (Figure 4.24).

10 What do we mean by a *renewable resource*? What, if anything, is a non-renewable resource? Give examples.

Figure 4.24 Chemical defences of bracken.

bracken leaves contain:

- an inhibitor of seed germination (released into the soil and leaf litter around the plant)
- an insect moulting hormone (interferes in growth of insects browsing on the plant)
- an enzyme destroying B vitamins in animals that eat the bracken
- a carcinogen active in cattle that feed on bracken and which they secrete in milk taken for human consumption

Bracken dramatically demonstrates that chemical defences exist in plant tissues. In fact a variety of chemical compounds, found in different species of plants, provide protection against other organisms. Some are highly toxic and kill predators when consumed in small doses, such as the alkaloids in rhododendrons. Others are cumulative in their effects, such as the tannins of oaks and many other plants. These are protein precipitants, making food less digestible and causing starvation and stunting.

Pyrethrum

Several relatives of the garden chrysanthemum plant produce an organic chemical in their tissues, known as a terpene – a natural insecticide. The presence of this chemical is harmless to the plant, but harms insects that attack the tissues. The active compound has been extracted and identified; it is named pyrethrum.

Pyrethrum is now a commercially exploitable example of targeted 'chemical warfare' initiated by plants. The ground-up fruits of *Chrysanthemum cinerariifolium*, known as the Dalmatian chrysanthemum, are a productive source. Pyrethrum and related substances (called pyrethrins) may damage the nervous system of insects, and when present in sub-lethal doses, are insect repellents, also. For example, some of the pyrethrum-producing plants are grown as 'companion plants' in rows beside vegetable crops that are vulnerable to damage by insect predation. Destruction by these pests is significantly reduced by this gardener's ploy.

Synthetic pyrethroids (based on natural pyrethrum) have been developed in an attempt to exceed the effectiveness of the naturally occurring form by enhancing stability in sunlight. Also, genetic engineers have sought to transfer the genes for this natural, endogenous insecticide to other plants.

11 There are obvious advantages of endogenous insecticides being present in valuable crop plants. What potential disadvantages exist, do you think?

Antimicrobial activity

There is some evidence that naturally occurring chemicals with antimicrobial properties are formed in plants, too. Table 4.3 lists examples.

Table 4.3 Naturally occurring antimicrobial chemicals from plants.

Plant source	Active agent(s)	Target micro-organisms
garlic (*Allium sativum*)	allicin, an ester formed from the amino acid alliin by the action of an enzyme, alliinase; substrate and enzyme only mix if the plant cells are crushed (or are damaged by predator activity)	allicin has been shown to be a potent antibiotic when tested on a spectrum of bacteria that have become resistant to many drugs, including MRSA
horse-mint (*Mentha longifolia*)	and other species of mint essential oils including menthol and carvone	these substances have been shown to be active as antibiotics against *Bacillus subtilis*, *Kocuria kristinae*, *Staphylococcus aureus* and *S. epidermidis*

Actually, most plant diseases due to micro-organisms are caused by parasitic fungi, rather than by bacteria. (The reverse is the case in animals – here fungal diseases are rare, and most animal diseases of this type are due to parasitic bacteria.)

Of course, plants are exposed to huge and diverse populations of micro-organisms in the soils they grow in, but much of plant life appears to grow strongly and healthily.

In fact, many plants live in close associations with soil micro-organisms to varying degrees of mutual benefit. There is a huge microflora immediately around root systems, a region known as the rhizosphere. Here, metabolites beneficial to micro-organisms may escape or be released from the roots. The resulting enhanced growth and metabolism of these micro-organisms may benefit the plant in either or both of the following ways.

- The micro-organisms make essential inorganic ions more readily available to the plant.
- The vigorous growth of the huge populations of harmless micro-organisms occurs at the expense of others, including pathogenic micro-organisms, the presence of which might otherwise harm the plant.

⊒L
www
A* Extension 4.3:
Defences must fail – in
the long term, at least!

Some soil fungi are actually in closer relationships with plant roots, in structures called mycorrhiza. The biology of these relationships is more complex.

■ How Science Works 4.3: Criterion 4 – Planning an investigation of the antimicrobial properties of plant extracts

This is a challenging project that will need careful preparation, followed by discussion with your tutor to refine the most appropriate way of proceeding in the time available.

Safety issues are of paramount importance when culturing micro-organisms.
You will need to decide on the plant material to be investigated, how a reproducible extract can be obtained, the cultures of bacteria that it is safe to use, how they can be cultured so that plant extracts can be applied, and exactly how the enquiry can be conducted safely in a manner that is most likely to give significant results.

Plant material
If you choose garlic, for example, you need to select a way of crushing tissue and obtaining an extract that can be investigated. This may be carried out in a pestle and mortar with a weighed quantity of tissue and a measured quantity of distilled water added. Will the extract need centrifuging to obtain a sample that allows the investigation to be replicated, for example?

Will the preparation of the extract itself need to be delayed until you have bacterial populations established in Petri dishes on which to test its effects?

Agar plate with bacterial 'lawn' to which extracts may be applied
You may remember demonstrations of investigations of the effects of various antibiotics on the growth of colonies of bacteria that cover the surface of an agar plate. This bacterial growth is called a 'lawn', and the antibiotics were present on specially prepared filter papers applied to the lawn surface. Adverse effects on the growth of the colonies became visible on subsequent incubation, as the bacterial cells were killed by potent antibiotics diffusing out from the filter paper discs. The outcomes were observed through the glass or plastic Petri dishes without their being opened (a safety requirement). This technique can be adapted for your enquiry here.

However, you need to research (using a lab manual of A level practicals, or a practical study guide resource, perhaps) how a bacterial lawn is prepared and incubated, prior to application of filter papers with your experimental extract. The cultures of bacteria it is safe to use include strains of *Escherichia coli*, *Bacillus subtilis* and *Micrococcus luteus*.

When you prepare small filter paper discs to deliver your plant extract to a lawn, you may choose to set up different dilutions, or perhaps apply extracts at different time intervals after extraction.

How can you record the results you observe of the impact of extracts on bacterial growth?

What controls would be appropriate to ensure any observed effects were due to the plant extract alone?

Finally, a risk assessment of the method selected needs to be conducted to ensure safe working throughout. Take advice on this.

Case study 4 – Drugs from plants

Plants are the oldest source of medicines – leaves, seeds or fruits, bark and roots of plants, shrubs and trees have been the prime ingredients of medical remedies for centuries. The gathering of herbs and the preparation of remedies was a domestic skill practised in families. Many of our wild plants retain common names that allude to their postulated benefits as 'folk medicines'. Doubtless, many (but not all) of the advantages of these treatments lay in the confidence people had in their (unproven) healing powers – we might call some of them **placebos**. (Today, a placebo is defined as a medicine or treatment prescribed for psychological benefit to the patient, rather than for any physiological effects.) Some plant remedies were certainly harmful. From the sixteenth century, tobacco enjoyed a high *medicinal* reputation – it was commended as a cure for headaches, toothache, chilblains and kidney stones, among many other ills!

On the other hand, significant discoveries were made. In 1763, the Revd Edmund Stone observed that willow bark (*Salix alba*) was the source of a remedy for 'ague' (an illness involving fever and shivering – probably malaria). Also, on giving an extract of willow bark to 50 patients with rheumatic fever, he reported satisfactory improvements. We now know the active ingredient in this bark has a similar effect to aspirin. While this discovery was reported to the Royal Society, it was ignored at the time.

A commercial side to herbal medicine developed – and it continues. You may still visit the Chelsea Physic Garden in London ('physic' refers to the science of healing). This second oldest botanical garden in the UK was founded as an apothecaries' garden in 1673 by the Worshipful Society of Apothecaries. An apothecary was a person who prepared and sold medicines. Many of today's drugs have a botanical origin – too many to list in total. Aspirin is a synthesised copy of a chemical found in plants. Digitalis, used for slowing heart beats, comes from foxgloves, and opium poppies are the source of the opiates codeine and morphine. A more recent discovery is that the bark of the Pacific yew (*Taxus brevifolia*) contains an anti-cancer drug, known as taxol.

The development of modern drug-testing routines

Today in the UK, new drugs and medicines have to be approved by a government agency, the Medicines and Healthcare products Regulatory Agency (MHRA), following a rigorous regime of critical trials that also requires the approval of the MHRA. Manufacturers and distributors are licensed by the MHRA, too. The financial cost of a product does not enter into the Agency's decision-making process; there are other official bodies that assist our National Health Service (NHS) to decide on the cost-effectiveness or otherwise of a drug or medicine.

You can learn more about the work of the MHRA by visiting their website, www.mhra.gov.uk. The clinical trials procedures are summarised in Table 4.4. These trials extend over a prolonged period; it takes very many years to develop a drug.

Clearly there have been huge advances in the critical examination of proposed medical remedies in the time since Edmund Stone made his apparent discovery of an aspirin-like substance, and his attempt to publish it.

How has this come about?

A significant milestone along this road was the work of a Birmingham doctor, William Withering, who experimented on the effects of extracts of the foxglove plant. This poisonous plant (Figure 4.25) commonly seen on acidic soils was already used in moderation in ancient folk remedies to treat the condition called 'dropsy' (now known as **oedema**, an issue discussed in Chapter 1, page 24).

How did Withering become involved?

William Withering (1741–99) was born in Wellington, Shropshire, where his father was an apothecary. After graduating from the Edinburgh Medical School, he worked as a physician at Stafford Infirmary until, in 1775, he was recommended to take over a Birmingham practice by none other than the illustrious Erasmus Darwin, grandfather of Charles Darwin. Here Withering regularly enjoyed the company of several distinguished Enlightenment scientists and industrialists at meetings of their Lunar Society, of which Erasmus was a leader. These pioneer

Table 4.4 Stages of the drug-testing regime.

Stage	Procedure
1 Pre-clinical stage of drug development	Involves extensive laboratory tests to ensure safety when delivered to humans, covering the composition of the drug, its manufacture, and how it will be safely delivered. The effect of the drug on cells and tissues (*in vitro* studies), and trials with animals – referred to as 'animal models' – are studied. Toxicology studies are required to identify possible risks to humans. Then, formulation of the drug in a form in which it can be administered – e.g. as nasal spray, or time-release capsule – is carried out.

If the drug now receives approval for the next stages, these will all involve humans, i.e. *in vivo* studies. Successful outcomes are needed at each of the following stages for the trial to be allowed to continue.

Stage	Procedure
2 Phase I clinical studies	A number of healthy volunteers, typically 20–100 people, receive the drug for a limited period, after being fully briefed on its intended effects. Investigations monitor the way the drug is metabolised in the body and what effects it has. This phase lasts for less than one year.
3 Phase II clinical studies	A larger group of volunteer patients *with the condition the drug is designed to treat* are selected, normally involving a group of several hundred. Both safety and effectiveness of the drug are the focus of these trials, which require from six months to three years. The patients are typically divided into two or three randomised groups, one receiving the drug, one receiving a placebo (and possibly one group receiving the current standard treatment). These trials are 'double blind' – neither patients or the administrators of the treatments know which group is receiving the placebo or the drug.
4 Phase III clinical studies	Finally, expanded versions of Phase II trials are now mounted, randomised and double blind, as before. Many hundreds of patients with the condition are involved, and the trials last for up to four years.

If the outcomes of the above have all been satisfactory, the drug may be approved for a marketing licence.

Post-approval studies	Issues such as new side-effects that have emerged, and the responses of different age groups to the drug, may need to be reviewed.

Figure 4.25 Foxglove (*Digitalis purpurea*) in flower (July).

thinkers were focused on the development and applications of scientific processes and scientific ideas; their meetings will surely have been mutually inspirational in the development of their various contributions.

Withering progressed in his career. He became physician at Birmingham General Hospital, but by 1783 he was suffering from TB. The obvious distress this disease caused him led to the (no-doubt well-intentioned) epigram 'the flower of Physick is Withering'. He retired, but was later elected Fellow of the Royal Society.

HSW 4.4: Criterion 11 – The role of the scientific community

The steps in Withering's work that are significant in the story of the development of the customs and practice of modern drug testing are as follows.

1   As a country doctor, Withering had heard of a herbal remedy for dropsy. (We now know that this painful condition is due to the accumulation of tissue fluid in the body, and is typically caused by heart and kidney problems that raise blood pressure. First the tissue fluid that fails to return to the blood circulation accumulates in the limbs, but slowly it fills the lungs, too, causing them to fail, leading to death.)

2   It was after a chance encounter with a particular, knowledgeable patient who had the symptoms of dropsy and who proposed to treat herself with a potion of several mixed herbs that included *Digitalis*, that he took particular note. He found she quickly recovered! He obtained details of the composition of the remedy from his patient, and correctly guessed which ingredient was effective.

3   Another patient with symptoms of dropsy and an irregular heart beat was treated with his own formulation of 'digitalis soup' and promptly recovered. Unfortunately, the next patient with the same symptoms was elderly. She was not helped, and he lost confidence in this line of enquiry, for the time being.

4   However, after moving to the General Hospital in Birmingham, Withering was now presented with many patients with dropsy, and he was in a position to study both the adverse side effects of administering variations of his digitalis soup, and the effects of varying the dosages administered.

5   He devised and pioneered a procedure of steadily increasing the dosage until the symptoms began to subside but at levels low enough for no side-effects to develop.

6   The results of these experiments (and his revolutionary methods of drug testing) were eventually recorded in a book, *A Treatise on the Foxglove*, in 1775. This book also publicised his refined version of the long-standing and popular herbal remedy for dropsy. It was from much later analyses that the active ingredient in *Digitalis* was shown to be a substance we call digitalis (sometimes known as digoxin or digitoxin).

HSW 4.5: Criterion 11 – The Royal Society, and the role of the scientific community in validating new knowledge and ensuring integrity

4.4 Biodiversity and the environment

There are vast numbers of living things in the world. The word 'biodiversity' is a contraction of 'biological diversity', and is the term we use for this abundance of different types or species.

Proportions of known species:

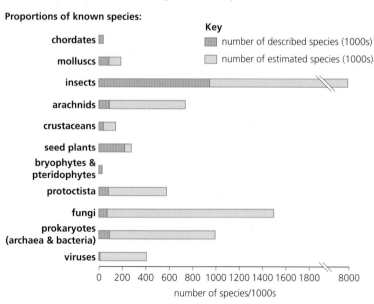

Key
- number of described species (1000s)
- number of estimated species (1000s)

Data from: 'The web of life; a strategy for systematic biology in the UK' The UK Systematic Forum, The Natural History Museum, Cromwell Road, London SW7 5BD

Relative numbers of animal and plant groups:

The 1.7 million described species are more than 50% insects, and the higher plants, mostly flowering plants, are the next largest group. By contrast, only 4000 species of mammals are known, about 0.25% of all known species.

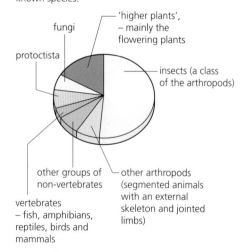

Figure 4.26 Introducing biodiversity.

Up to now, about 1.7 million species have been described and named. However, until very recently there has been no attempt to produce an international 'library of living things' where new discoveries are automatically checked out (see Extension below). Consequently, some known organisms may have been 'discovered' more than once.

Meanwhile, previously unknown species are being discovered all the time. In the UK alone, several hundred new species have been described in the past decade. We might have expected all the wildlife in these islands to be known, since Britain was one of the countries to pioneer the systematic study of plants and animals. Apparently, this is not the case; previously unknown organisms are frequently found here, too.

Worldwide, the number of unknown species is estimated at 3–5 million at the very least, and possibly as high as 100 million. So scientists are not certain just how many different types of organisms exist. Figure 4.26 is a representation of the proportions of known and unknown species estimated to exist in many of the major divisions of living things.

■ Extension: Species 2000 Programme

The Species 2000 Programme is part of a federation of databases and the organisations that are creating them, and was set up in 1994. It has the goal of creating a valid checklist of all of the world's species of plants, animals, fungi and micro-organisms. This huge task is to be achieved by bringing together existing species databases covering each of the major groups of organisms. The UK arm of this initiative is based at the University of Reading. Here, Species 2000 is in partnership with the Integrated Taxonomic Information System (ITIS) of North America, currently producing the Catalogue of Life, which is linked to the Global Biodiversity Information Facility (GBIF). You can keep up-to-date with developments at www.sp2000.org

What do we mean by 'species'?

By the term 'species', we refer scientifically to a particular type of living thing. A species is defined simply as 'a group of individuals of common ancestry that closely resemble each other, and are normally capable of interbreeding to produce fertile offspring'.

There are two issues about this definition to mention. One is that some (very successful) species reproduce asexually, without any interbreeding at all. Organisms that reproduce asexually are very similar in structure, showing little variation between individuals.

The other point is that we now believe that species change with time, and that new species evolve from other species. The fact that species may change makes it impossible to think of species as constant and unchanging. However, evolutionary change takes place over a long period of time. On a day-by-day basis, the term 'species' is satisfactory and useful.

Taxonomy – the classification of diversity

Classification is essential to biology because there are too many different living things to sort out and compare unless they are organised into manageable categories. With an effective classification system in use, it is easier to organise our ideas about organisms and make generalisations. The scheme of classification has to be flexible, allowing newly discovered living organisms to be added into the scheme where they fit best. It should also include fossils as these are discovered, since we believe living and extinct species are related.

The process of classification involves:

■ giving every organism an agreed name
■ imposing a scheme upon the diversity of living things.

The binomial system of naming

Many organisms have local names, but these often differ from locality to locality around the world, so they do not allow observers to be confident they are talking about the same thing. For example, in North America the name 'robin' refers to a bird the size of the European blackbird – altogether a different bird from the European robin (Figure 4.27).

Figure 4.27 Two 'robins' of the Northern Hemisphere.

European robin ♂
Erithacus rubecula
length 14 cm

American robin ♂
Turdus migratorius
length 21 cm

Instead, scientists use an international approach called the **binomial system** (meaning 'a two-part name'). By this system everyone, anywhere knows exactly which organism is being referred to. In the binomial system, each organism is given a scientific name consisting of two words in Latin. The first (a noun) designates the **genus**, the second (an adjective) the **species**. The generic name comes first, and begins with a capital letter, followed by the specific name. Conventionally, this name is written in *italics* (or is underlined). As shown in Figure 4.28, closely related organisms have the same generic name; only their species names differ. You will see that when organisms are frequently referred to the full name is given initially, but thereafter the generic name is shortened to the first (capital) letter. Thus, in continuing references to humans in an article or scientific paper, *Homo sapiens* would become *H. sapiens*.

12 Scientific names of organisms are often difficult to pronounce or even to remember. Explain why they are used.

Figure 4.28 Naming organisms by the binomial system.

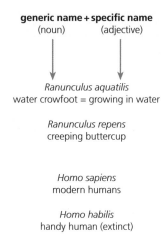

generic name + specific name
(noun)   (adjective)

Ranunculus aquatilis
water crowfoot = growing in water

Ranunculus repens
creeping buttercup

Homo sapiens
modern humans

Homo habilis
handy human (extinct)

The scheme of classification

The science of classification is called **taxonomy**. The word comes from 'taxa' (singular taxon), which is the general name for groups or categories within a classification system. The taxa used in taxonomy are given in Figure 4.29.

Biological classification schemes are the invention of biologists, based upon the best available evidence at the time. In classification, the aim is to use as many characteristics as possible in placing similar organisms together and dissimilar ones apart. Just as similar **species** are grouped together into the same **genus** (plural genera), so too, similar genera are grouped together into **families**. This approach is extended from families to **orders**, then **classes**, **phyla** and **kingdoms**. This is the hierarchical scheme of classification, each successive group containing more and more different kinds of organism.

Activity 4.7: Carl Linnaeus of Sweden, and the Linnean Society of London

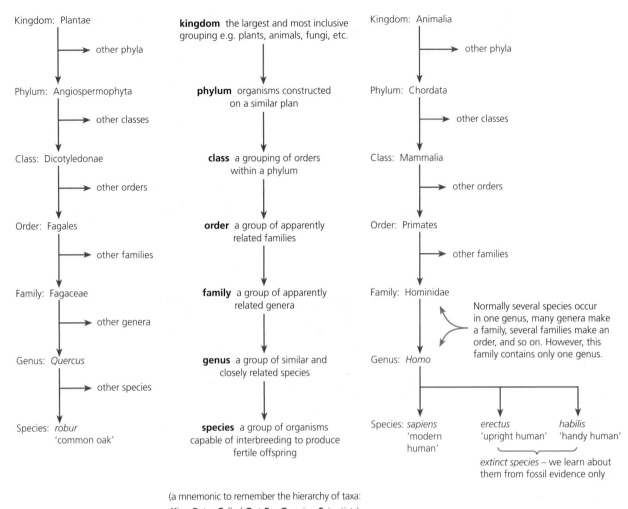

Figure 4.29 Taxa used in taxonomy, applied to genera from two different kingdoms.

Biodiversity in the environment

'Environment' is a term we commonly use for 'surroundings'. In biology, we talk about the environment of cells in an organism, or the environment of organisms in a habitat. It is our term for the external conditions affecting the existence of organisms, too, so 'environment' is a rather general, unspecific term, but useful, nonetheless.

We will now consider where 'diversity' is to be found within the environment, and aspects of how it can be measured.

'Hotspots' and endemism

When the distribution of livings things across the Earth's surface is investigated, we find that most species are not distributed widely at all. Instead, very many are restricted to a narrow range of the Earth's surface. The places where the majority of species occur are referred to as 'biodiversity hotspots'.

Visit www.biodiversityhotspots.org for an interactive map of the locality of hotspots around the world. Figure 4.30 is a representation of part of the Earth's land mass; it shows species distribution in North and South America. You can see that very many of the biodiversity hotspots are in the tropics. As a consequence, when tropical rainforests are destroyed, the only habitat of a huge range of plants is lost, and with them very many of the vertebrates and non-vertebrates dependent on them, too. This is one major reason why the fate of the remaining rainforest is such an urgent issue. We will return to this issue and its consequences for conservation of biodiversity later in this chapter.

Figure 4.30 Most species have small ranges and these are concentrated unevenly.

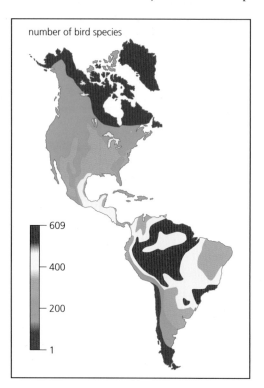

number of bird species

609
400
200
1

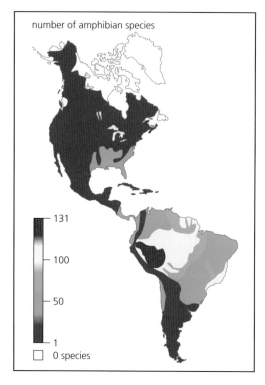

number of amphibian species

131
100
50
1
☐ 0 species

A unique form of geographic isolation is the phenomenon of 'endemism'. **Endemic** is a term used for a particular plant or animal taxon that is *always* restricted to a narrow geographical range. In this region, these organisms occur, and they are not to be found anywhere else.

What environmental processes may be responsible for these special cases of what is, in effect, geographical isolation?

We can identify three causative processes underlying endemism:

1   The phenomenon of plate tectonics is occasionally responsible for the isolated evolution of a limited number of species.
    Movements of the tectonic plates may lead to an area of land – part of a continent, in effect – becoming completely isolated from the larger land mass, as the plates continue to 'drift' apart. Following isolation, the plant and animal life marooned on the new 'island' evolves quite differently from that on the mainland. Examples include the island of Madagascar that broke away from the mainland of Africa in geological history. The unique lemurs of this island are particularly beautiful (and vulnerable) examples of endemic species, but the majority of the plants, breeding birds, amphibians and reptiles now present on Madagascar are endemic species, too.

2   Islands formed by volcanic activity are sterile outcrops of rocks, some in ocean settings, well away from mainland masses. Weathering generates soil there, and plant life becomes established when seeds are blown in or existing vegetation is detached from elsewhere and happens to be 'rafted' across to the island. Bird and insect life may fly or be carried in on the wind, and many terrestrial animals other than mammals may well survive extended, accidental 'raft' journeys on fallen tree trunks that happen to reach the new shore, and become introduced by chance. With the passage of time, the species that arrive and survive may evolve into species unique to the island. For example, many of the species now found on the Galapagos Islands are endemic there, including many plant species and iguanas.

3   'Islands' created within existing land masses may exceptionally be sites for the evolution of new, endemic species. An example is the isolated lakes that formed in the rift valley in Africa. These now support endemic fish populations.

A* Extension 4.4:
Primary and secondary
successions – an outline

Habitat diversity

Just as with huge areas of the Earth's surface, so with habitats we find that the distribution of species is often uneven. Activity 4.8 concerns the community of organisms in sand dunes as they age, from pioneering dunes that sea and wind are currently depositing and which plant life has hardly stabilised as yet, to a mature stable dune community with some invasions of plants from other habitats.

Early on in the development of this succession, the number and diversity of species is low. At this stage, the populations of organisms present are usually dominated by **abiotic factors**. An abiotic factor is a non-biological factor (for example, temperature), that is part of the environment of an organism. For example, if a prolonged period of very low temperatures occurs in a habitat, the number of organisms may be severely reduced.

On the other hand, in a stable climax community many different species are present, many in quite large numbers. In this situation, adverse abiotic conditions are less likely to have a dramatic effect on the numbers of organisms present. In fact, in a well-established community, the dominant plants (a **biotic factor**) set the way of life for many other inhabitants. They provide the nutrients, determine the habitats that exist, and influence the environmental conditions. These plants are likely to modify and reduce the effects of extreme abiotic conditions, too.

Species richness and biodiversity

Species richness is defined as the total number of different species within a given area or community. To produce this information, a precise listing of all the different types of organisms is required. However, the abundance of each species present is not required. As such, species richness is not a complete measure of biodiversity of a habitat. The diversity of species present in a habitat may be measured by applying the formula known as the Simpson Diversity Index:

13 In a vegetable plot
left fallow (left
unsown, typically
for a year), three
weed species
appeared. Individual
plants were
counted.

$$\text{diversity} = \frac{N(N-1)}{\Sigma n(n-1)}$$

where N = total number of organisms of all species found,
and n = number of individuals of each species.

Make yourself familiar with this formula by answering self-assessment question 13, followed by Activity 4.8.

Species	n
groundsel	45
shepherd's purse	40
dandelion	10
total (N)	95

Calculate the
Simpson Diversity
Index for this
habitat.

Diversity within a species – variety of alleles within a gene pool

Individual organisms of the same species have many features in common – that is how we are able to identify them. But these same individuals also show significant differences, one from another.

How does diversity within a species arise?

The processes underlying genetic diversity have already been discussed (Chapter 2 – mutations, page 118 – and Chapter 3 – variations arising in meiosis and fertilisation, page 122). Review these sources and respond to self-assessment question 14.

Activity 4.8:
Calculating the Simpson
Diversity Index for a
sand-dune community

A* Extension 4.5:
Understanding 'abiotic'
and 'biotic' factors

14 Design a flow chart that records in sequence the sources of variation that may arise, starting with the genotype of a parent and leading to the distinctive phenotype of one of its offspring.

Measuring genetic variability

In a population – a group of individuals of a species living close together and able to interbreed – the alleles of the genes located in the reproductive cells of those individuals make up a **gene pool**. A sample of the alleles of the gene pool will contribute to form the genomes (gene sets of individuals) of the next generation, and so on, from generation to generation.

The size of an interbreeding population has a direct impact on the genetic diversity of the individuals. A very small population can be described as an **inbreeding** group – the individuals are closely related. In fact, the smaller the population, the more closely related the offspring will be. The important genetic consequence of inbreeding is that it leads to **homozygosity** – at more and more of the loci there will be homologous alleles. There is progressively less variation in the population. While the individuals of that population may initially be well adapted, in the face of environmental changes the population is less able to adapt. The genetic fitness of a population is compromised.

It is in these latter cases of small, isolated populations that genetic variability is most critical. In the cases of populations of endangered species, the question is whether there is sufficient genetic diversity to allow the population to adapt to future changes in the environment and so survive.

This practical problem faces modern zoos, which have taken on a role of attempting to conserve genetic variation within endangered species via captive breeding programmes.

It is also an issue for endangered organisms in the wild, where population numbers have been reduced to small, isolated groups in former strongholds. This problem can be illustrated by examining current conservation research being undertaken in the tropical rainforests of Sabah in Malaysia, Borneo.

A conservation case study based on genetic diversity analyses

Orang-utans (*Pongo pygmaeus*), with gorillas and chimpanzees, are great apes. Most of their features (including their large cranium and well-developed brains, elongated arms and highly developed muscles) are common to humans too, although they are not our direct ancestors. Orang-utans are the largest ape species after the gorilla – females typically weigh in excess of 35 kg and males up to 80 kg. A standing male may be up to 150 cm high. They are arboreal, active by day, with a diet largely of fruits and shoots. They may be solitary, or live in pairs, or very small family groups, sleeping in tree 'nests', and moving on from day to day. The young (birth weight about 1.5 kg) remain with their mother for up to five years of parental care. They reach maturity at about 10 years, and their total life expectancy is 30 years or more. Humans are their only predator, either directly (through poaching) or indirectly (through logging and forest clearing for agriculture).

Sabah, situated on the north-eastern part of the island of Borneo, has an orang-utan population that has been in decline for the past 100 years (Table 4.5). This collapse was triggered by massive deforestations begun in the 1890s and accelerated in the 1950s and 1970s.

Table 4.5 Estimated orang-utan numbers in Sabah, Borneo, over the past 100 years or so.

Year	Number of orang-utans
1900	310 000
1980	25 000
2003	13 000

Now, the orang-utans of Sabah are threatened with extinction in the near future, due to the continuing logging of the forest trees, clearing of whole forests for oil palm plantations, and illegal killings. Today's population estimates are based on ground surveys and are confirmed by aerial surveys (nest counts) made by helicopter. While Sabah is the main stronghold of these primates in North Borneo, the present estimated population is of only 11 000 individuals, distributed in the remaining pockets of rainforest, as shown in Figure 4.31. More than half live outside protected areas, in forests still frequently disturbed by selective logging activities.

A team of conservation biologists, led by Professor Mike Bruford and Dr Benoit Goossens of the Biodiversity and Ecological Processes group, School of Biosciences, Cardiff University, are investigating the genetic diversity of these populations in an attempt to devise sustainable schemes that will effectively support remaining orang-utan populations. Their data are being collected from animals living in Kinabatangan Wildlife Sanctuary (Figure 4.31). By collecting hair and faeces found at fresh nest sites, they are able to extract DNA to create genetic profiles. Genetic markers called microsatellites are applied to the DNA samples which distinguish, among other things, heterozygous genes from homozygous genes. They serve as tools to evaluate inbreeding levels, the genetic structure and past history of populations, and to assess both gene flow between populations and effective population size. (The techniques of DNA extraction and analysis that underpin DNA profiling, for example, are introduced in outline in Figure 2.54, page 93, but otherwise comprise part of the A2 Biology programme.)

The results establish that, in Kinabatangan Sanctuary, the orang-utans do not cross the rivers but do move freely through the forest areas on either side. Studies indicate that, if nothing is done about the existing sanctuary provision here, then this situation will almost certainly lead to extinction. If an elaborate series of wildlife corridors are set up between remaining areas of forest, survival can hopefully be assured. Since these measures are probably more expensive than local economies can sustain, the team has devised a programme of transfers of individuals between sites (referred to as translocations), together with a modest programme of corridor establishment. This proposal, known as the Kinabatangan Management Plan, is being prepared for presentation to the agencies and authorities who must find the necessary funding for whichever measures they choose to support.

15 What are the potential consequences for the orang-utan populations of Sabah of the development of eco-tourism provisions there?

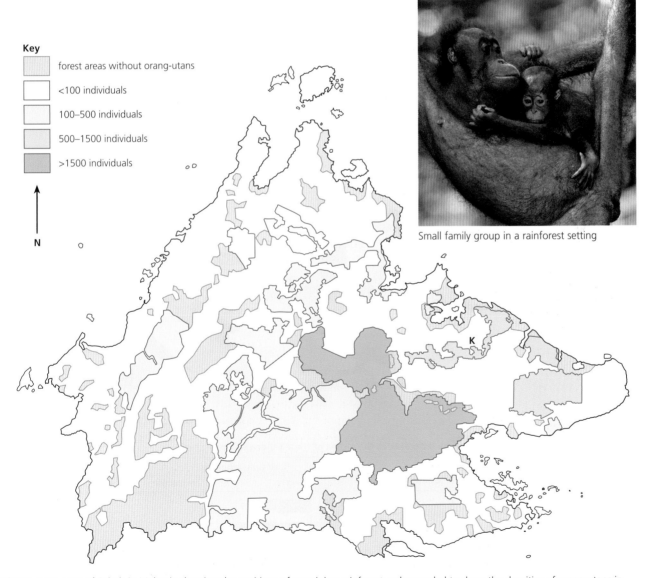

Small family group in a rainforest setting

Key

- forest areas without orang-utans
- <100 individuals
- 100–500 individuals
- 500–1500 individuals
- >1500 individuals

N

Figure 4.31 Map of Sabah in Malaysia showing the positions of remaining rainforest, colour coded to show the densities of orang-utans in each; the Kinabatangan Wildlife Sanctuary is marked **K**.

Niche – a concept central to ecology

Ecology is the study of living things within their environment. It is an essential part of modern biology – understanding the relationships between organisms and their environment is just as important as knowing about the structure and physiology of animals and plants, for example. One of the ideas that ecologists have introduced into biology is that of the 'ecosystem'. An **ecosystem** is defined as a community of organisms and their surroundings, the environment in which they live. An ecosystem is a basic functional unit of ecology since the organisms that make up a community cannot realistically be considered independently of their physical environment. An example of an ecosystem is woodland.

Within an ecosystem are numerous habitats. The term **habitat** refers to the place where an organism lives. Within a woodland ecosystem, for example, some organisms have a habitat restricted to a small area. A leaf-tissue parasite such as the holly leaf-miner insect, especially at the larval stage, is an example, for it is restricted to the interior of the holly leaf. Other species are found very widely, such as *Pleurococcus*, the single-celled alga found on all damp surfaces, such as most tree trunks and branches. So, there is no particularly precise definition of a habitat – but the term is useful.

On the other hand, the term **niche** is more informative. It defines just how an organism feeds, where it lives, and how it behaves in relation to other organisms in its habitat. A niche identifies the precise conditions a species needs.

We can illustrate the value of the niche concept by reference to the two common and rather similar sea birds, the cormorant and the shag (Figure 4.32). Both birds live and feed along the coastline, and they rear their young on similar cliffs and rock systems. We can say that they apparently share the same habitat. However, their diet and behaviour differ. The cormorant feeds close to the shore on sea-bed fish, such as flatfish. The shag builds its nest on much narrower cliff ledges. It also feeds further out to sea, and captures fish and eels from the upper

Figure 4.32 The sea birds cormorant and shag – their niches.

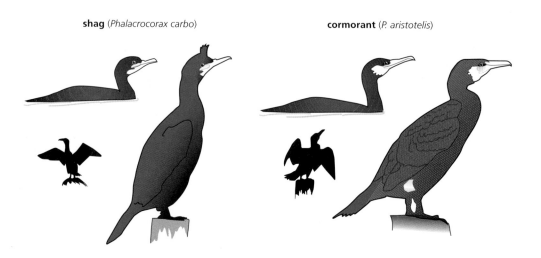

shag (*Phalacrocorax carbo*)

cormorant (*P. aristotelis*)

diet is a key difference in the niches of these otherwise similar birds

prey		% of prey taken by	
		shag	cormorant
surface-swimming prey	sand eels	33	0
	herring	49	1
bottom-feeding prey	flatfish	1	26
	shrimps, prawns	2	33

A* Extension 4.6:
'Niche' and the
competitive exclusion
principle

layers of the waters. Since these birds feed differently and have different behaviour patterns, although they occur in close proximity, they avoid **competition**. They have different niches.

Competition refers to the interaction between two organisms striving for the same resource. For competing organisms, the more their habitats overlap, the more both will strive to secure a finite resource. Hence potential competitors have evolved different niches.

Adaptation of organism to environment

Adaptation is the process by which an organism becomes fitted to its environment. There are countless examples of this process to be observed in all habitats.

Physiological adaptation

An example of a physiological adaptation is shown in the marine algae known as 'greens', 'browns' and 'reds', which flourish at different zones of the shoreline community (Figure 4.33). The colour differences in these seaweeds are due to the particular photosynthetic pigments they contain. These pigments enable algae to absorb and exploit different wavelengths of light. In the marine environment, with increasing depth, progressively more of the higher wavelengths of white light are absorbed or scattered by the sea water and its suspended particles. Consequently, the red algae, equipped to absorb the blue–green light that is transmitted to greater depths, flourish there. Here, the brown and green algae cannot photosynthesise because the wavelengths of light they are adapted to absorb do not reach that depth.

Meanwhile, at lesser depths, red seaweeds are progressively crowded out in competition with the vigorous growing brown seaweeds and green seaweeds as their particular pigments permit the efficient absorption of the incident light available, closer to the surface.

Incidentally, the barnacles *Chthamalus* and *Semibalanus* exhibit differing abilities to endure exposure in the intertidal zone, and this too is an example of physiological adaptation.

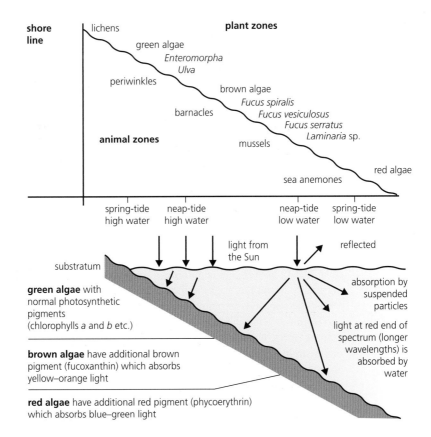

Figure 4.33 Zonation of green, brown and red seaweeds.

Behavioural adaptation

An example of a behavioural adaptation is the fighting behaviour of male marina iguana (*Amblyrhynchus cristatus*) of the Galapagos Islands (Figure 4.34). Fighting between members of the same species is almost universal among vertebrate species. This trait serves the important function of 'spacing out' individuals into non-overlapping territories, large enough to support an animal, its mate, and their offspring. Overcrowding is prevented. Fighting also results from competition for a mate, resulting in breeding by the stronger, more determined males. Consequently, fighting is as common among herbivorous animals as among carnivores – in fact, the marine iguana is a herbivore, grazing on the seaweed fronds that clothe the rocks below the tide line.

However, fighting between individuals of the same species rarely ends in death or even injury – fights are highly ritualised. Harm is prevented when the loser (the weaker or younger or less determined combatant) adopts a submission posture, signalling that defeat is accepted. At the earliest opportunity, a retreat is affected, and injury to males of the same species is avoided.

Incidentally, the differing feeding habits of the cormorant and shag (Figure 4.32) are also examples of behavioural adaptation.

Figure 4.34 Ritualised combat to maintain territory among marine iguanas.

Arrival and recognition of an intruding male in the territory of another male.

Territory is defended by lunging at the intruder. Heads clash, and the intruder is driven back.

When the intruder realises that defeat is inevitable he drops into a submission posture. The conflict is terminated, and neither animal is injured.

Anatomical adaptations

Examples of anatomical adaptations are body structure adapted to regulate heat loss in various mammals. Mammals are described as **endotherms**, since their body's heat comes from the metabolic reactions of many body organs. Body temperature is largely regulated by varying heat loss from the body. The total heat produced from internal organs largely depends upon the volume of the body, but heat loss is dependent upon the surface area. Now, as the size of an organism increases, the volume increases more rapidly than the surface area. Consequently, animals in cold regions of the world tend to be large animals. An example is the polar bear. Smaller animals in colder regions need a high metabolic rate, and consequently require a regular and substantial food supply to survive.

Meanwhile, mammals of hot regions typically have external ears adapted as efficient radiators.

These flaps of skin bear little fur (hair – a heat insulation layer), and are richly supplied with blood capillaries. Heat brought to the external ears from the body interior by warm blood is quickly lost when capillaries here are dilated. A comparison of ear size in hares and rabbits in natural habitats at various latitudes on the North American continent appears to support this (Figure 4.35).

Figure 4.35 External ear size of hares in relation to latitude.

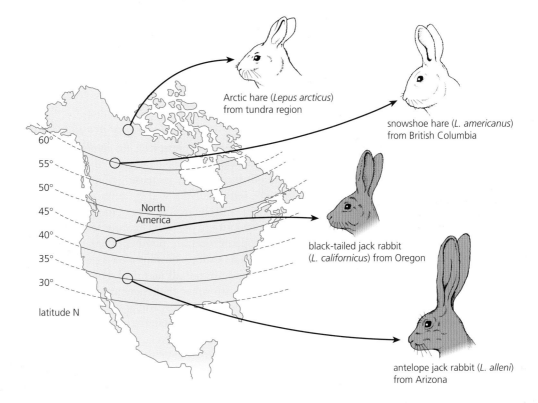

How adaptation is brought about – natural selection

It was Charles Darwin whose careful observations over many years led him to realise what natural process brought about the adaptations of organisms in response to challenging environmental conditions. He coined the term 'natural selection' for this.

Charles Darwin published these ideas in 1859, in a book titled *The Origin of Species by means of Natural Selection*, published by John Murray of Albemarle Street, London. He was proposing a mechanism for evolution of organisms.

By 'evolution', we mean the gradual development of life in geological time. The word evolution is used very widely, but in biology it specifically means the processes by which life has been changed from its earliest beginnings to the diversity of organisms we know about today, living and extinct. It is the development of new types of living organisms from pre-existing types by the accumulation of genetic differences over long periods of time.

Explaining natural selection today

Charles Darwin (and nearly everyone else in the scientific community of his time) knew nothing of Mendel's work. Instead, biologists generally subscribed to the concept of 'blending inheritance' when mating occurred (which would only reduce the genetic variation available for natural selection, if it actually occurred). Today we are really talking about 'Neo-Darwinism' which is essentially a restatement of the concepts of evolution by natural selection in terms of Mendelian and post-Mendelian genetics.

The evidence and arguments for natural selection are as follows.

1 **Organisms produce many more offspring than survive to be mature individuals**. Darwin did not coin the phrase 'struggle for existence', but it does sum up the point that the over-

production of offspring in the wild leads naturally to their competition for resources. Table 4.6 is a list of the normal rates of production of offspring in some common species.

How many of these offspring survive to breed, themselves?

Table 4.6 Numbers of offspring produced.

Organism	Number of eggs/seeds/young per brood or season
rabbit	8–12
great tit	10
cod	2–20 million
honey bee (queen)	120 000
poppy	6000

In fact, in a stable population, a breeding pair may give rise to a single breeding pair of offspring, on average. All their other offspring are casualties of the 'struggle'; many organisms die before they can reproduce.

So, populations do not show rapidly increasing numbers in most habitats, or at least, not for long. Population size is naturally limited by restraints we call 'environmental factors'. These include space, light, and the availability of food. The never-ending competition for resources results in the majority of organisms failing to survive and reproduce. In effect, the environment can only support a certain number of organisms, and the number of individuals in a population remains more or less constant over a period of time.

2 **The individuals in a species are not all identical**, but show variations in their characteristics. Today, modern genetics has shown us that there are several ways by which genetic variations arise in gamete formation during meiosis, and at fertilisation. Genetic variations arise via:

- **random assortment** of paternal and maternal chromosomes in meiosis – this occurs in the process of gamete formation (page 118)
- **crossing over** of segments of individual maternal and paternal homologous chromosomes that results in new combinations of genes on the chromosomes of the haploid gametes produced by meiosis (page 116)
- the **random fusion of male and female gametes** in sexual reproduction – this source of variation *was* understood in Darwin's time.

Additionally, variation arises due to **mutations** – either chromosome mutations or gene mutations (page 79).

As a result of all these, the individual offspring of parents are not identical. Rather, they show variations in their characteristics.

3 **Natural selection results in offspring with favourable characteristics**. When genetic variation has arisen in organisms:

- the favourable characteristics are expressed in the phenotypes of some of the offspring
- these offspring may be better able to survive and reproduce in a particular environment; of course, others of the offspring will be less able to compete successfully, survive and reproduce.

Thus natural selection operates, determining the survivors and the genes that are perpetuated in future progeny. In time, this selection process leads to adaptation to environment, later to new varieties and then to new species.

The operation of natural selection is sometimes summarised in the phrase 'survival of the fittest', although these were not words that Darwin used, at least not initially. To avoid the criticism that 'survival of the fittest' is a circular phrase (how can fitness be judged except in terms of survival?), the term 'fittest' is applied in a particular context. For example, the fittest of the wildebeest of the African savannah (hunted herbivore) may be those with the acutest senses, quickest reflexes, and strongest leg muscles for efficient escape from predators. By natural selection, the health and survival of wildebeests is assured.

16 Explain the importance of modern genetics to the theory of the origin of species by natural selection.

■ Extension: Charles Darwin – the person

Charles Darwin (1809–1882) was a careful observer and naturalist who made many discoveries in biology. After attempting to become a doctor (at Edinburgh University), and then a clergyman (at Cambridge University), he became the unpaid naturalist on an Admiralty-commissioned expedition to the southern hemisphere, on a ship called *HMS Beagle*. On this five-year expedition around the world, and in his later investigations and reading, he developed the idea of organic evolution by natural selection.

Darwin remained very anxious (always) about how the idea of evolution might be received, and he made no moves to publish it until the same idea was presented to him in a letter by another biologist and traveller, Alfred Russel Wallace. Only then did he publish.

HSW 4.6: Criterion 7 – The tentative nature of scientific knowledge

Environmental change and adaptation – a topical example

In the face of environmental change, individuals possessing a particular allele or combination of alleles may be more likely than others to survive, breed, and pass on their alleles. The current predicament for our National Health Service of multiple antibiotic resistance in bacteria is an example of this situation (Figure 4.36).

Figure 4.36 Multiple antibiotic resistance in bacteria.

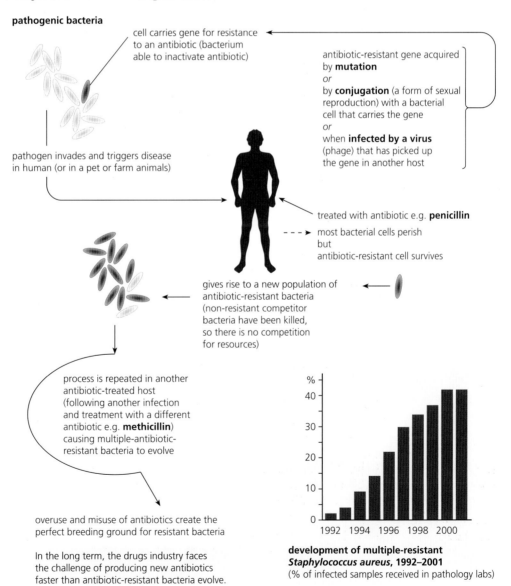

pathogenic bacteria

cell carries gene for resistance to an antibiotic (bacterium able to inactivate antibiotic)

antibiotic-resistant gene acquired by **mutation**
or
by **conjugation** (a form of sexual reproduction) with a bacterial cell that carries the gene
or
when **infected by a virus** (phage) that has picked up the gene in another host

pathogen invades and triggers disease in human (or in a pet or farm animals)

treated with antibiotic e.g. **penicillin**

- - - ▸ most bacterial cells perish but antibiotic-resistant cell survives

gives rise to a new population of antibiotic-resistant bacteria (non-resistant competitor bacteria have been killed, so there is no competition for resources)

process is repeated in another antibiotic-treated host (following another infection and treatment with a different antibiotic e.g. **methicillin**) causing multiple-antibiotic-resistant bacteria to evolve

overuse and misuse of antibiotics create the perfect breeding ground for resistant bacteria

In the long term, the drugs industry faces the challenge of producing new antibiotics faster than antibiotic-resistant bacteria evolve.

development of multiple-resistant *Staphylococcus aureus*, 1992–2001
(% of infected samples received in pathology labs)

17 Explain:
 a why doctors ask patients to ensure that they complete their course of antibiotics fully
 b why the medical profession tries to combat resistance by regularly alternating the type of antibiotic used against an infection.

A* Extension 4.7: The development of the ideas of evolution

Certain bacteria cause disease, and patients with bacterial infections are frequently treated with an antibiotic to help them overcome the infection. Antibiotics are very widely used.

In a large population of a species of bacteria, some may carry a gene for resistance to the antibiotic in question. Sometimes such a gene arises by spontaneous mutation; sometimes the gene is acquired in a form of sexual reproduction between bacteria of different populations.

The resistant bacteria in the population have no selective advantage in the absence of the antibiotic, and must compete for resources with non-resistant bacteria. But when the antibiotic is present, most bacteria of the population will be killed off. Then the resistant bacteria are very likely to survive and be the basis of the future population. In the new population, all now carry the gene for resistance to the antibiotic; the genome has been changed abruptly. Incidentally, one can speculate as to whether this is a case of natural selection or artificial selection – but the principles are the same!

Biodiversity and taxonomy

Schemes of classification of the organisms of the living world are devised by interested humans, using the best available evidence. Schemes are frequently revised, based on critical evaluation of new data and evidence. For example, at one time the living world seemed to divide naturally into two kingdoms:

- **plants** – photosynthetic organisms (autotrophic nutrition), mostly rooted (that is, stationary)
- **animals** – organisms that ingest complex food (heterotrophic nutrition), typically mobile.

Later, new evidence arrived with the application of electron microscopy to biology (page 99). Examination of EMs of cells of plants, animals, bacteria and other groups led to the discovery of the two types of cell structure, namely **prokaryotic** and **eukaryotic**. As a result, the division of living things into two kingdoms needed overhauling. Today, the agreed division of living things is into five kingdoms, as listed on the next page.

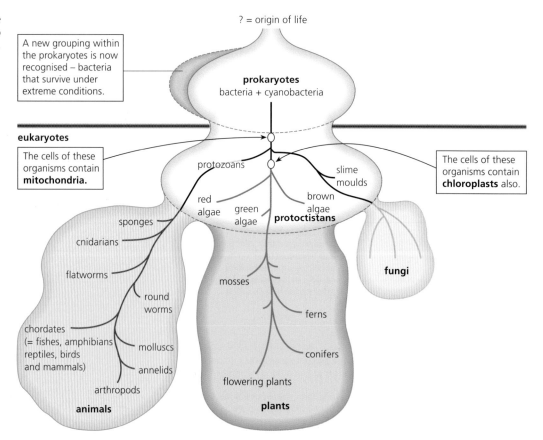

Figure 4.37 Possible evolutionary relationship of the five kingdoms.

18 Why are viruses hard to include in the living or non-living categories of matter, yet fungi are clearly defined as a kingdom of living things?

1 **Prokaryotae** – the prokaryote kingdom, the bacteria and cyanobacteria (photosynthetic bacteria), predominantly unicellular organisms
2 **Protoctista** – the protoctistan kingdom (eukaryotes), predominantly unicellular, and seen as resembling the ancestors of the fungi, plants and animals
3 **Fungi** – the fungal kingdom (eukaryotes), predominantly multicellular organisms, non-motile, and with heterotrophic nutrition
4 **Plantae** – the plant kingdom (eukaryotes), multicellular organisms, non-motile, with autotrophic nutrition
5 **Animalia** – the animal kingdom (eukaryotes), multicellular organisms, motile, with heterotrophic nutrition.

New taxonomic groupings – the three domains

Vast numbers of micro-organisms occur in the biosphere immediately around us, and many of them exist in environmental conditions similar to those of plants and animals. However, new and continuing studies of life under extremes of environmental conditions – including extremes of temperature or salinity, and of pressure at great ocean depths – have led to the discovery of a huge and growing range of previously unknown micro-organisms. Now, a systematic hunt for life forms in 'extreme' places (known as **extremophiles**) is underway. The biochemistry of the cells of extremophiles has proved distinctly different from that of existing known life forms, too.

The numbers and different types of known extremophiles are increasing as the importance of these strange habitats becomes better understood, but they include micro-organisms that:

■ are salt-loving (halophiles), found in salt lakes and where sea water has become concentrated and salt has crystallised
■ require extremely alkaline conditions above pH10 (alkalinophiles), found in soda lakes
■ survive abnormally high temperatures (thermophiles)

These evolutionary relationships have been established by comparing the sequence of bases (nucleotides) in the ribosomal RNA (rRNA) present in species of each group.

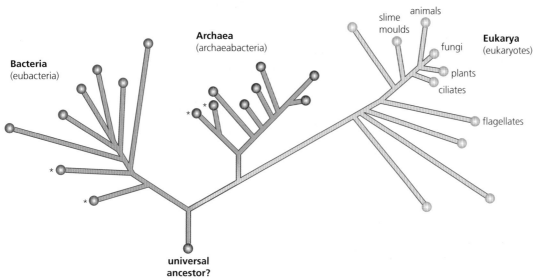

* The shortest branches lead to hyperthermophilic species which suggests that the universal ancestor of all living things was a hyperthermophile (possibly 'assembled' under conditions at deep ocean vents where volcanic gases are discharged into water at high temperature and pressure).

Figure 4.38 The classification of living organisms into domains.

- withstand 250 atmospheres pressure (barophiles), as well as extremely high temperatures
- survive extremely cold habitats (psychrophiles).

The micro-organisms of extreme habitats have cells that we can identify as prokaryotic. That is, there is no true nucleus present, nor is the cytoplasm packed with the range of organelles so typical of eukaryotic cells. However, the biochemistry of the cells of extremophiles has proved distinctly different from existing known life forms. In particular, it is the larger RNA molecules present in the ribosomes that have proved to be tell-tale molecules.

How are the RNA molecules of extremophiles analysed?

First, ribosomes (sites of protein synthesis) are isolated from the cytoplasm of cells, and their RNA is extracted. Then the sequence of nucleotides present (that is, the sequence of the bases cytosine, guanine, adenine and uracil in this RNA) is determined. Finally, comparisons are made between the base sequences in these RNA molecules with those from the previously known and classified prokaryotes, and with RNA from eukaryotes. The outcome has been the discovery of new evolutionary relationships, and the development of a new scheme of classification of living things. We now recognise *three* major forms of life, and these are called **domains**.

The organisms of each domain share a distinctive, unique pattern of ribosomal RNA which establishes their close evolutionary relationship. These domains are:

- the Archaea or archaeabacteria (the extremophile prokaryotes)
- the Bacteria or eubacteria (the true bacteria)
- the Eukarya (all eukaryotic cells – the protoctists, fungi, plants and animals).

The characteristics of the three domains

The archaeabacteria are organisms about which we are least familiar, so we concentrate on their unique features first, before we can compare their structure with that of cells of organisms of the other two domains (Table 4.7).

The archaeabacteria are a diverse group in both structure and physiology. Some are single cells (Figure 4.39), others are filamentous or other forms of aggregates of cells. We have noted that many are found in extreme conditions (the extremophiles); some live in terrestrial habitats, others are aquatic. For example, archaeabacteria represent about a third of all prokaryotic biomass in polar coastal waters. Others live symbiotically in the gut of animal species – as do vast numbers of true bacteria, too.

19 Discovery of prokaryotic organisation depended on the development of the electron microscope. On the development and application of what additional techniques was the discovery of the 'domain' level of organisation dependent?

Figure 4.39 A typical methanogenic bacterium of the Archaea.

Methanobrevibacter ruminantium

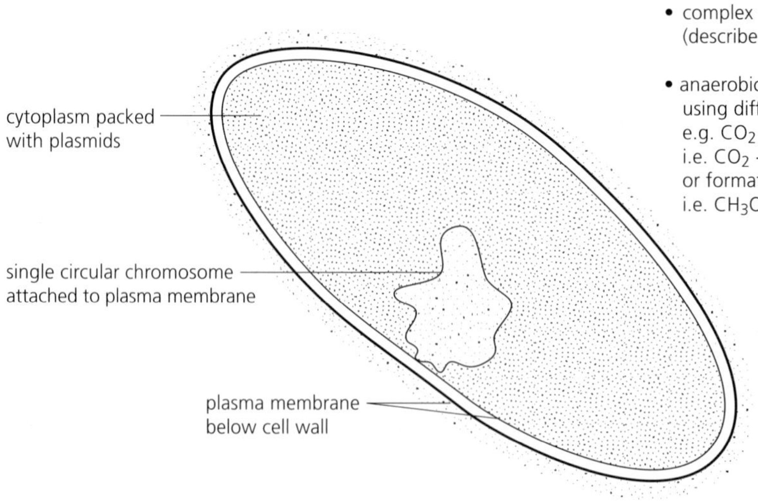

Features:

- short, rod-shaped bacterium

- complex wall chemistry (described as pseudopeptoglycan)

- anaerobic respiration, using different substrates, e.g. $CO_2 + H_2$
  i.e. $CO_2 + H_2 \rightarrow CH_4 + 2H_2O + energy$
  or formate (methanoate)
  i.e. $CH_3OH + H_2 \rightarrow CH_4 + H_2O + energy$

cytoplasm packed with plasmids

single circular chromosome attached to plasma membrane

plasma membrane below cell wall

(cell approximately 0.7 μm in diameter)

The distinguishing features of the archaeabacteria are:

- **the chemistry of their cell walls is different** from the chemistry of the walls of the true bacteria. Often a variety of complex polysaccharides are present, but *none* contains the peptidoglycan or particular amino acids characteristic of the eubacteria. Some have a substantial layer of protein or glycoprotein external to the polysaccharides, and some have walls of protein only.
- the lipids of the cell membranes found in the Archaeabacteria are **glycolipids that have branched-chain hydrocarbons attached to glycerol** (in place of the fatty acids of the lipids of the eubacteria and eukaryotes). Polar lipids are also present, and the whole combine in different ways to form membranes of variable thickness and rigidity.
- they have a single circular chromosome, without **introns** (non-coding sequences of nucleotides, page 131), as the eubacteria do. However, they posses **far fewer genes** than the true bacteria. Also, their nucleic acid is extremely variable in the proportions of the two bases, guanine (G) and cytosine (C), present. In species where the nucleic acid has been sequenced, it is found that the genes are unlike those in true bacteria and eukaryotes. Also, there are relatively few plasmids present.
- the **ribosomes** are small (70S), like those of the eubacteria, but they are more **variable in shape and chemistry**. It is the biochemical analysis of their ribosomes upon which the evolutionary relationships of the members of the three domains is based.
- their physiology, metabolism and lifestyles are extremely variable, when compared with members of the other two domains.

Clearly, a range of evidence, based on developments in analytical and investigative techniques, have been exploited in the development of this latest taxonomic grouping. For the purposes of comparison, the structure of a typical bacterium is shown in Figure 3.11 (page 106), and that of a human liver cell – typical of a eukaryotic cell – in Figure 3.4 (page 103).
Look those up again now, to help you make sense of the table below.

Table 4.7 The cellular characteristics of the three domains.

	Archaeabacterial cell e.g. Figure 4.39	Eubacterial cell e.g. Figure 3.11	Eukaryotic cell e.g. Figure 3.4
ribosomes – size and chemical analysis	70S size – unique biochemistry	70S size – unique biochemistry	80S – unique biochemistry
histones with the nucleic acid of chromosome(s)	absent	absent	present, forming nucleosomes
introns within chromosome(s)	absent	absent	present, between exons
plasmids in cytoplasm	few present	many present	typically absent
cell walls – present or absent, and chemistry	present – chemically different from those of eubacteria	present – chemically different from those of Archaea	present in green plant cells (cellulose) and in fungi (chitin), but absent in animal cells
cell membranes – composition of lipids	contain glycolipids with hydrocarbon chains, in plasma membrane	contain lipids with fatty acids, in plasma membrane	contain lipids with fatty acids, in all membranes present

Biodiversity and conservation

New species evolve, but other species (less suited to their environment, perhaps) become extinct, as much of the fossil record throughout geological time indicates. One example of a well-documented extinction is that of the dodo bird.

The extinction of the dodos

The dodo was an inhabitant of the island of Mauritius in the Indian Ocean. It was a bird related to modern pigeons. Over geological time, this distinctive organism had evolved to master a

Figure 4.40 A reconstruction of the dodo bird.

terrestrial habit. In the process, it became a large bird (it was about a metre long and had a mass of approximately 20 kg).

The dodo nested on the ground, and reared its young there. The diet was one of seeds and fruits that had fallen from the forest trees. It was one of many forest-dwelling birds on the island – one of the 45 species for which there are early records, of which only 21 species have survived to this day.

Why have these well-adapted organisms died out?

Factors that contributed to the extinction of the dodos

Mauritius is a medium-sized island, extremely far from any mainland. The dodo's misfortune was that Mauritius got involved in the spice trade. Since ancient times, spices have been brought to European centres from the East, where they are grown and produced. The rarity of spices in the West, and their importance at times when food was difficult to preserve fresh, gave them great value. In order to break the monopoly in spice trading held by the people of Venice from 1200 to 1500, other European explorers from western European sea-going countries sailed the sea-routes from early in the sixteenth century. They made 'stop-over' visits to ports and islands to restock (fresh water and fresh meat were required). Visits lead to population migrations, and immigrant peoples brought their farm stock with them. Mauritius was a popular location for this.

Records show the dodo became extinct on Mauritius by 1681. The specific factors that contributed to this were as follows.

- **Visiting sailors killed the dodo as a source of fresh meat** to restock their ships. The dodo proved an easy victim, unfamiliar with humans (or other mammals) – it had previously lacked significant natural enemies.
- Later, **settlers brought cats, dogs and pigs**, and inadvertently, **rats** – all alien species which fed on the occupants of the dodos' nests.
- **Natural habitats on the island were deliberately destroyed** as land was cleared for agriculture.

The impact of human population growth – increasing extinctions

The current human population explosion began at about the beginning of the Industrial Revolution, some 200 years ago, and it continues still. No one is certain when the rate of population growth will slow down, as it surely must. When we plot world human population against time, we see a steeply rising J-shaped curve – our population growth appears to be in an exponential (log) phase of growth.

Figure 4.41 The changing pattern of the estimated world human population.

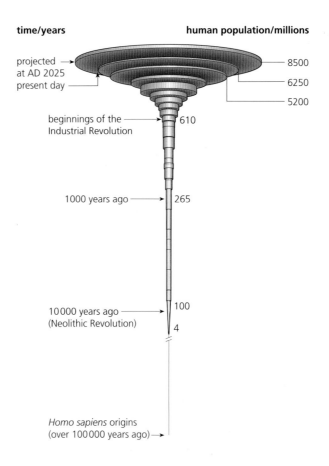

20 In what ways does the human population explosion represent a threat to the Earth's wildlife?

A* Extension 4.8: Introducing the biomes of the world

Today, the impact of humans on the environment is very great indeed – human activities are changing biomes worldwide, and in far-reaching ways. A **biome** is a major life-zone over a large area of the Earth, characterised by the dominant plant life present. Examples include rainforests and grasslands (savannah). There is virtually no part of the biosphere which has not come under human influence and been changed. Change continues – in both the physical and chemical (abiotic) environment, and to the living organisms (the biota). Many of these changes threaten biodiversity. We can illustrate this using the case of the rainforests.

The case of the vanishing rainforests

Rainforests cover almost 2% of the Earth's land surface, but they provide habitats for almost 50% of all living species. It has been predicted that if all non-vertebrates occurring in a single cubic metre of tropical rainforest soil were collected for identification, there would be present at least one completely previously unknown species. Tropical rainforests contain the greatest diversity of life of any of the world's biomes.

Sadly, tropical rainforests are being rapidly destroyed. Satellite imaging of the Earth's surface provides the evidence that this is so – where no other reliable sources of information are available. The world's three remaining tropical forests of real size are in South America (around the Amazon Basin), in West Africa (around the Congo Basin) and in the Far East (particularly but not exclusively on the islands of Indonesia).

The current rate of destruction is estimated to be about one hectare (100 m × 100 m – a little larger than a football pitch) every second. This means that each year an area larger than the British Isles (31 million hectares) is cleared. While extinction is a natural process, this current rate is on a scale equivalent to that at the time of the extinction of the dinosaurs (an event 65 million years ago, at the Cretaceous–Tertiary boundary).

Extinctions have always been a feature of the history of life, but we should strive to conserve biodiversity in this ecosystem for the reasons set out in Table 4.8.

Table 4.8 Why conserve rainforest ecosystems?

Ecological reasons	Most species of living things are not distributed widely, but instead are restricted to a narrow range of the Earth's surface. In fact, the majority of species living today occur in the tropics (Figure 4.30). So when tropical rainforests are destroyed, the only habitat of a huge range of plants is lost, and with them very many of the vertebrates and non-vertebrates dependent upon them.
	In effect, the rainforests are critically important 'outdoor laboratories' where we learn about the range of life that has evolved, the majority of which consists of organisms as yet unknown.
	The soils under rainforest are mostly poor soils that cannot support an alternative biota for very long. To destroy rainforest is to remove the most productive biome on the Earth's surface – in terms of converting the Sun's energy to biomass.
	If destroyed but later left to regenerate, only species that have not become extinct may return. Re-grown rainforest will be deprived of its variety of life.
Economic reasons	The whole range of living things is functionally a gene pool resource, and when a species becomes extinct its genes are permanently lost. The destruction of rainforests decreases our genetic heritage more dramatically than the destruction of any other biome. As a consequence future genetic engineers and plant and animal breeders are deprived of a potential source of genes. Many new drugs and other natural products, in some form or another, are manufactured by plants. The discovery of new, useful substances often starts with rare, exotic or recently discovered species.
	7 million km² of humid tropical forests have so far been cleared – about half of the original forest present before clearance programmes started. Much is 'cleared' to make way for the production of crops for food (Figure 4.42). Yet only 2 million km² have remained in agricultural production. Mostly, the soil does not sustain continued cropping.
	Rainforest is a continuing resource of hardwood timber, which if selectively logged can be productive, but when cleared as forest, is lost as a source of timber in future.
	Trees help stabilise land and prevent disastrous flooding downriver. Huge areas of productive land are washed away, once mountain rainforest has been removed.
	Trees in general are carbon dioxide 'sinks' that help reduce global warming. Without these trees, other ways of reducing atmospheric CO_2 must be found.
Aesthetic reasons	These habitats are beautiful, exhilarating and inspirational places to visit. They are part of the inheritance of future generations which should be secured for future people's enjoyment, too.
Ethical reasons	This biome is home to many forest peoples who have a right to traditional ways of life.
	Similarly, many higher mammals, including relatively close 'relatives' of *Homo*, live exclusively in these habitats. The needs of all primates must be respected.

Activity 4.9: The Red List of Threatened Species

Red Data Books as listings of indicator species

Environmentalists seek the survival of endangered species by initiating and maintaining local, national and international scientific research programmes to identify critical issues in all the world's biomes. The International Union for the Conservation of Nature and Nature Reserves co-ordinates the updating of Red Data Books. These list endangered species, identifying those for which special conservation efforts are needed. The health and general wellbeing of populations of these organisms are indicators of environmental changes.

Practical conservation – what does it entail?

Conservation involves applying the principles of ecology to manage the environment so that, despite human activities, a balance is maintained. The aims of conservation are to preserve and promote habitats and wildlife, and to ensure natural resources are used in a way that provides a sustainable yield. Conservation is an active process, not simply a case of preservation (Table 4.9), and there are many different approaches to it. Practical conservation involves:

Figure 4.42 Rainforest land for agriculture!

Adapted from: *THE TIMES* Thursday February 2 2006

Rainforest Eden faces death by chainsaw – to make margarine

A £4.4bn plantation could destroy an ecological treasure, reports **Nick Meo** from Borneo.

IN THE cool of dawn Betung Kerihun could almost be an English wood until the honks of a rhinoceros hornbill echo around the great creeper-festooned trees.

The forest is an almost untouched Eden. Orang-utans and gibbons live high in the canopy. On the forest floor clouded leopards and eight-metre pythons hunt wild boar and deer and are themselves hunted by one of the last truly nomadic forest peoples, the Penan. But this rainforest, which has survived for millions of years, may now be doomed.

A £4.2 billion plan proposed by a Chinese bank and backed by Jakarta politicians would clear 1.8 million hectares of this wilderness over the next six years to grow oil palms to feed the world's growing appetite for margarine, ice-cream, biscuits and biodiesel fuel.

Until now Betung Kerihun, technically a protected national park, has been saved by its remoteness despite the network of roads that illegal loggers have begun to push inside its boundaries.

What survives is a biological treasure that staggers scientists newly arrived from Europe. It is home to thousands of tree frogs, bats and orchids. More than 1,000 insects have been identified in a single tree. In one ten-year period 361 new plant, animal and insect species were discovered in Borneo.

Charles Darwin, who explored the giant island before writing *The Origin of Species*, called it "one great untidy luxuriant hothouse made by nature for herself".

Reaching the virgin jungle from the coast takes days by river or logging road across a landscape of tree stumps, and the most common sound is that of the chainsaw. Only half of the island is now covered with forest, compared with three quarters in the 1980s.

Conservationists believe that the real motive for the oil palm scheme is to gain access to the valuable timber along a great swath of the Indonesian-Malaysian border.

"It's a scam," said Stuart Chapman, of the WWF conservation group. "Palm oil is a lowland equatorial crop and not suited to steep upland soils. There are two million hectares of idle land, already cleared of forest, in lowland Borneo which is suitable for planting. but putting this plantation in the island's centre would give logging interests the excuse they need to cut down trees."

- the designation and maintenance of representative habitats as nature reserves
- preservation of endangered species and their genetic diversity through the maintenance of botanical and zoological gardens (with their captive breeding programmes), and the establishment of viable seed banks.

We look into these issues next.

Conservation by promotion of nature reserves

Nature reserves comprise carefully selected land set aside for restricted access and controlled use, to allow the maintenance of biodiversity, locally. This is not a new idea; the New Forest in southern Britain was set aside for hunting by royalty over 900 years ago. An incidental effect was to produce a sanctuary for wildlife.

Today, this solution to extinction pressures on wildlife includes the setting up and maintenance of areas of special scientific interest as nature reserves, of our National Parks (the first National Park was Yellowstone, set up in North America in 1872), and of the African game parks which have been more recently established. In total, these sites represent habitats of many different descriptions, in many countries around the world. Some of the conservation work they achieve may be carried out by volunteers.

What biogeographical features of a reserve best promote conservation?

In a nature reserve, the area enclosed is important – a tiny area may be too small to be effective. The actual dimensions of an effective reserve vary with species size and life style of the majority of the threatened species it is designed to protect, however.

Also, for a given reserve there is an 'edge effect'. A compact reserve with minimal perimeter is less effective than one with an extensive perimeter interface with its surroundings. The use to which the surrounding area is put is important, too; if it is managed sympathetically, it may indirectly support the reserve's wildlife.

Another feature is geographical isolation – reserves positioned at great distances from other protected areas are less effective than reserves in closer proximity. Also, it has been found that connecting corridors of land are advantageous. In agricultural areas these may simply take the form of hedgerows protected from contact with pesticide treatments that nearby crops receive.

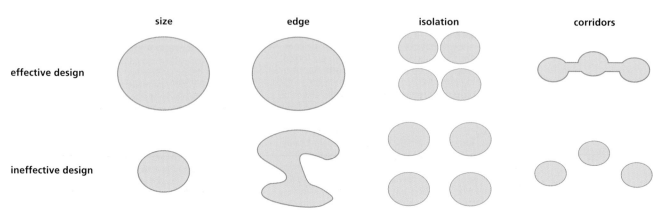

Figure 4.43 Features of effective nature reserves.

Table 4.9 What active management of nature reserves involves.

Continuous monitoring of the reserve so that causes of change are understood, change may be anticipated, and measures taken early enough to adjust conditions without disruption, should this be necessary.

Maintenance of effective boundaries, and the limiting of unhelpful human interference. The enthusiastic involvement of the local human community communicates messages about the purposes of conservation (a local 'education' programme, in effect), and that everyone has a part to play.

Measures to facilitate the successful completion of life cycles of any endangered species for which the reserve is 'home', together with supportive conditions for vulnerable and rare species, too.

Restocking and re-introductions of once-common species from stocks produced by captive breeding programmes of zoological and botanical gardens.

21 Evaluate the particular challenges involved in the management of a nature reserve near your home, school or college.

Ex-situ conservation – an appraisal

Zoological (and botanical) gardens and their captive breeding programmes

Endangered species typically have very low population numbers and are in serious danger of becoming extinct. For some species whose numbers have dwindled drastically, captive breeding may be their last hope of survival.

Today, many zoos co-operate to manage individuals of the same species, held in different zoos, as a single population. A 'stud book' – a computerised database of genetic and demographic data – has been compiled for many of these species. This provides the basis for the recommendation and conduct of crosses designed to preserve the gene pool and avoid the problems of inbreeding.

Animals may be shipped between zoos, or the technique of artificial insemination may be employed. Some species can be very hard to breed in captivity, while with others there has been a high success rate.

The final success is when individuals bred in captivity are released and have survived in the wild. Examples include red wolves, Andean condors, bald eagles and golden lion tamarins.

Criticism of the process includes the fear that genetic diversity may have dwindled so much already that a species cannot be regenerated, that the work concentrates on a few, highly attractive species, and that at great cost (with funds diverted from more effective habitat conservation) it gives a false sense that extinction problems are being solved.

The building up of seed banks

Storing seeds in seed banks is an inexpensive and space-efficient method of *ex-situ* conservation (Figure 4.44). The natural dormancy of seeds allows for their suspended preservation for long periods, typically in conditions of low humidity and low temperature. The steps, following collection and preparation are:

- seed drying (to below 7% water)
- packaging (in moisture-proof containers)
- storage (at a temperature of $-18\,°C$)
- periodic germination tests
- re-storage or replacement.

The advantages of all these approaches are reviewed in Table 4.10, below.

Table 4.10 Pros and cons of *in-situ* and *ex-situ* conservation.

In-situ conservation **terrestrial and aquatic nature reserves**	*Ex-situ* conservation **captive breeding programmes of zoological and botanical gardens, and seed banks**
Habitats that are already rare are especially vulnerable to natural disaster – rare habitats themselves are easily lost, if a range of examples are not preserved as nature reserves.	Originally, zoos were collections of largely unfamiliar animals kept for curiosity, with little concern for any stress caused, but now captive breeding programmes make good use of these resources.
When a habitat disappears the whole community is lost, threatening to increase total numbers of endangered species.	Captive breeding maintains the genetic stock of rare and endangered species.
A refuge for endangered wildlife allows these species to lead natural lives in a familiar environment for which they are adapted, and be a part of their normal food chains.	The genetic problems arising from individual zoos having very limited numbers to act as parents is overcome by inter-zoo co-operation (and artificial insemination in some cases).
The biota of a reserve may be monitored for early warning of any further deterioration in numbers of a threatened species, so that remedial steps can be taken.	Animals in zoos tend to have significantly longer life-expectancies, and are available to participate in breeding programmes for much longer than wild animals.
The offspring of endangered species are nurtured in their natural environment and gain all the experiences this normally brings, including the acquisition of skills from parents and peers around them.	Captive breeding programmes, for most species they are applied to, have been highly successful, although the young do not grow up in the 'wild', so there is less opportunity to observe and learn from parents and peers.
There is an established tradition of maintaining reserves and protected areas in various parts of the world, so there is much experience to share on how to manage them successfully	Captive breeding programmes generate healthy individuals in good numbers for attempts at re-introduction of endangered species to natural habitats – a particularly challenging process, given that natural predators abound in these locations.
Nature reserves are popular sites for the public to visit (in approved ways), thereby maintaining public awareness of the environmental crisis due to extinctions, and individual responsibilities that arise from it.	Zoos and botanical gardens are accessible sites for the public to visit (often sited in urban settings where many may have access), contributing effectively to public education on the environmental crisis.
Reserves are ideal venues to which to return endangered individuals which are the product of captive breeding programmes – providing realistic conditions for re-adaptation to their habitat, where progress can be monitored.	Seed banks are a convenient and efficient way of maintaining genetic material of endangered plants, which make use of the ways in which seeds survive long periods in nature.

Figure **4.44** Wellcome Trust Millennium Building, Wakehurst Place – home to the Millennium Seed Bank (an initiative of the Royal Botanic Gardens, Kew).

4 End-of-topic test

(full End-of-topic tests are available on the DL Student website)

1 a Identify the organelle with each of the roles or functions listed below.
   i site where amino acids are assembled into proteins
   ii structure regulating exit and entry to a large permanent vacuole
   iii small spherical sac containing hydrolytic enzyme
   iv site of carbon dioxide fixation and oxygen gas formation
   v site of mRNA formation
   b List any of the organelles you have identified in part **a** above that are formed or contained by a double membrane. (7)

2 The following is an account of plant cell walls. Read the text and then fill in the blanks, using the most appropriate term from this list:
impervious; plasmodesmata; meshwork; fibres; microfibres; lignin; secondary; middle lamella; primary; water; cytokinesis; cellulose

When a growing plant cell divides on completion of mitosis, a cell wall is laid down across the old cell, dividing the contents, as part of _____.

In the first step, a gel-like layer of calcium pectate, known as the _____, is delivered and deposited by vesicles cut-off from the Golgi apparatus and endoplasmic reticulum. These vesicles coalesce along the midline of the dividing cell. Some of the endoplasmic reticulum of the parent cell becomes trapped across this layer at various points. These persist and form the cytoplasmic connections between the new cells, called _____.

Onto this early wall material is added several layers of cellulose microfibres, typically deposited transversely. They form the _____ cell wall. Then, as the cell enlarges, microfibres begin to slide past each other as they re-orientate in response to the strains imposed by the lengthening cell. A _____ of microfibres results.

Subsequently, more layers of _____ are normally added to the inner surface of the wall, forming _____ cell wall. These additional layers are mostly added after growth in overall size of the cell nears completion. The combined effect of layers of _____, lying at different angles, is to increase the strength of the wall. In cells clearly specialised for support, the secondary layers become very thick indeed – much of the interior of the cell may be taken up by wall material. This is the case in the plant cells called _____.

In addition, certain cells' walls may become impregnated with a complex hydrophobic material called _____ which further strengthens them. It is the xylem vessels and fibres that have this addition. Their walls are now more-or-less _____. Consequently, the pits that form in xylem vessel walls (and are found in many other plant cell walls, too) are the main channels by which _____ moves from xylem to the living cells of root, stem and leaves. (12)

3 In an investigation of ion uptake, bean and maize plants were maintained with their roots in culture solutions under aerobic conditions, for a period of four days. The initial concentration of ions in the culture solution is recorded below, together with the concentrations of the same ions in the root tissue at the end of the period of ion accumulation.

Ion	Initial concentration in the medium/mM	Concentration in root tissue after 4 days/mM	
		maize	bean
K^+	2.0	160	84
Ca^{2+}	1.0	3	10
Na^+	0.32	0.6	6
P_i*	0.25	6	12
NO^{3-}	2.0	38	35
SO_4^{2-}	0.67	14	6

*P_i = inorganic phosphate, which has several ionic forms

data from *The Physiology of Flowering Plants* by H. Opik and S. Rolfe (2005) CUP, page 119

a The culture solution was kept vigorously aerated throughout the experiment. What was the significance of this condition for plant root cells?

b Comment on the concentrations of potassium and nitrate ions recorded above, compared with those of the other ions listed.

c Taken as a whole, what does this data suggest to you about the mechanism of ion uptake in plant roots generally?

d At the same time, uptake of water will have occurred from the culture solution into the plants. Contrast the mechanisms of water uptake and ion uptake. (12)

4 Imagine that a new drug proposed for the treatment of prostate cancer had successfully completed the pre-clinical stage of drug testing at the laboratory of the pharmaceutical firm that discovered it, and that the outcome was promising at this stage.

a Outline the procedures of the subsequent three phases of clinical trials that would be required under UK legislation.

b Identify the main ethical issue that a thoroughly conducted clinical trial raises for the medical profession and society. (10)

5 By means of succinct definitions and appropriate examples, explain what you understand by the following terms:
a biodiversity
b endemism
c species richness
d genetic diversity. (12)

6 By reference to examples, explain the concepts of 'habitat' and 'niche'. (8)

7 a Explain how the force that carries water up the stem of a plant in the xylem is generated.
b Identify the structural features by which a xylem vessel is adapted for its role as conduit for water transport.
c What properties of water make possible its movement through the plant in this way?
d Describe the ways in which the structures of fibres and xylem vessels differ. (12)

8 a What structures are common to animal and plant cells, when viewed by electron microscopy?
b Complete the table by listing further differences in structure between plant and animal cells. (12)

Plant cells	Animal cells
cellulose cell wall present	no cell wall present

9 a By means of fully annotated diagrams, describe the structure of cellulose.
b Starch is also a polysaccharide made from glucose monomers. List *three* ways in which the structure of starch differs from that of cellulose. (12)

10 a Outline the biochemical evidence on which the existence of three major forms of life called 'domains' was based.
b Complete the blank boxes in the table of characteristics of the three domains. (8)

	Archaeabacteria	Eubacteria	Eukaryote
ribosomes – size	70S		80S
histones with the nucleic acid of chromosomes		absent	present
introns within chromosomes	absent	absent	
plasmids in cytoplasm		many present	

11 a Outline *three* reasons in support of our striving to maintain species diversity in the face of increasing extinctions, today.
b Explain by means of examples the difference between in-situ and ex-situ conservation. (12)

Answers to SAQs

1 Lifestyle, health and risk

1 See Student DL Disk *Appendix 1 – Background chemistry for biologists*: sections entitled *Elements, atoms, molecules and compounds*, *How atoms form molecules* and *Ions*.

2 In a solution of glucose, water is the solvent and glucose is the solute.

3 The nucleus of a hydrogen atom, which is what remains when the atom loses its single electron, consists of a single proton.

4 Carbon dioxide and carbonates such as calcium carbonate.

5 See Student DL Disk *Appendix 1 – Background chemistry for biologists*: section entitled *How atoms form molecules*.

6 The presence of sugar in the cytoplasm or vacuole of a cell has a powerful osmotic effect (see *Osmosis – a special case of diffusion*, page 50). On the other hand, if glucose molecules are condensed to starch (or glycogen, in animal cells), these carbohydrate reserves are insoluble, and have no effect on cell water relations. Meanwhile, starch and glycogen may be speedily hydrolysed to sugar if and when the need arises.

7 A polymer is a large organic molecule made of repeating subunits known as monomers, chemically combined together. Among the carbohydrates, starch, glycogen and cellulose are all polymers built from a huge number of molecules of glucose (the monomer, combined together in different ways in these three, distinctive polymers).

8 See *Condensation and hydrolysis reactions*, page 6.

9 Blood cannot be directed to a respiratory surface immediately before or after servicing tissues that are metabolically active (and so have a high rate of respiration). Rather, the blood circulates randomly around the blood spaces and blood vessels.

10 clotting factor release → thrombin formation → fibrin formation

11 These non-elastic strands keep the heart valve flaps pointing in the direction of the blood flow. They stop the valves turning inside out when the pressure rises abruptly within the ventricles.

12 a Pressure in the aorta is always significantly higher than in the atria because blood is pumped under high pressure into the aorta, and during diastole and atrial systole the semilunar valves prevent backflow from the aorta. Meanwhile, blood enters the atria under low pressure from the veins, and the pumping action of the atria is slight compared to that of the ventricles, which generates pulse.

  b Pressure falls abruptly in the atrium once ventricular systole is underway because atrial diastole begins then.

  c The semilunar valve in the aorta opens when pressure in the left ventricle exceeds pressure in the aorta.

  d When ventricular diastole commences, pressure in the ventricles starts to fall and the bicuspid valve opens when the ventricular pressure falls below atrial pressure.

  e About 50% of the cardiac cycle is given over to diastole – the resting phase in each heart beat. The heart beats throughout life and takes limited rest at these moments.

13 Red cells, platelets and many of the blood proteins are not found in tissue fluid.

14 Endothelium lining of artery damaged (e.g. due to fatty deposits beneath it, effect of high blood pressure, effect of components in cigarette smoke) → platelets stick to artery wall at damaged site → blood clot formation triggered (see Figure 1.17) → thrombus becomes circulating embolus.

15 Advantage: Enables blood clotting cascade of events to be triggered at site of leak in circulation system, leading to repair of the blood vessel.

Disadvantage: Danger of thrombus formation, leading to a stroke or heart attack.

16 Because two events (A and B) regularly occur together, it may appear to us that A causes B. This is not necessarily the case. In fact there may be a common event that causes both, for example, or it may be an entirely spurious correlation.

In order to prove a causal relationship in science:
- a body of data about the possible relationship is collected and subjected to statistical tests
- where these tests indicate the probability of a direct link, then the nature of a possible linkage is investigated. For example, a persistent condition of hypertension is correlated with a raised incidence of coronary heart disease and 'vascular accidents'. In this case, the statistical relationship has been followed up, enabling us to understand why hypertension has these effects. Once the connections between events or conditions are understood, the relationship has been established.

So, a correlation does not prove the cause, but this does not mean there cannot be a causal relationship. A correlation provides statistical confidence in the possibility of a causal link, and as such is a springboard to further investigation, rather than proof of a relationship.

  a In this case, the text says that excessive dietary salt *causes* raised blood pressure, and we know that experimental evidence exists to demonstrate *how* raised blood pressure causes heart disease. So, from the information in the text, we can say that the association between heart disease and diets high in salt is a causal one.

  b On the other hand, the text says that in people with diets persistently low in antioxidants, there is evidence of higher incidences of heart disease, but does not provide any information beyond this correlation. So, we cannot assume a causal link – only the suggestion of a connection, which could be investigated further to see if direct causation between the two variables exists.

17 a To drive the circulation of blood through arteries, capillaries and veins, back to the heart.

  b See *Hypertension*, page 26.

18 The probability of the outcome of the dice landing 'six' face up is:

$$P = \frac{\text{number of nominated outcomes}}{\text{total number of possible outcomes}}$$

$$= \frac{1 \text{ (the 'face' with six dots)}}{6 \text{ (number of possible faces)}}$$

$P = 1$ in 6, or 0.166 666 recurring
$= $ approximately 0.17 or 17%

19 'Risk' is the probability of the occurrence of an unwanted event. In 2005, total deaths due to CVD were $2597 + 1643 = 4240$ per million of the population.

$$\text{risk} = \frac{1 \text{ in } 1\,000\,000}{4240}$$

$$= 1 \text{ in } 235.8$$

$$= \frac{1}{235.8}$$

$= 0.004240$ or 0.42%

20 A control group in a medical investigation:
- comprises the same number of people as a patient group selected for an experimental treatment

- is as comparable with the patient group as possible, as regards the ages, genders, health, occupations, and general life experiences of its members
- consists of people having the same 'condition' for which treatment is being given to the experimental group.

The fates of the control group and patient group are followed in comparable detail in order to evaluate the efficacy of the experimental treatment. Note that the use of such a control group means that a treatment is withheld from people who might have benefited. This raises obvious ethical issues.

21 When yours and your peers' results are to hand, it may be valuable to share, examine and discuss them. What general conclusions emerge?

22 The outer layers of the artery and arteriole (and vein) are strengthened by fibrous connective tissue that contains collagen fibres. Healthy arteries resist the wave of pressure (pulse) with each heart beat; arteries with weakened walls are more likely to be damaged by blood pressure.

2 Cells, genes and health

1 The essential processes characteristic of living things are:
- transfer of energy (respiration)
- feeding or nutrition
- metabolism
- excretion
- movement and locomotion
- responsiveness or sensitivity
- reproduction
- growth and development.

2 $1\,mm = 1000\,m$

$$\frac{1000}{100} = 10$$

So 10 cells of $100\,m$ diameter will fit along $1\,mm$.

3

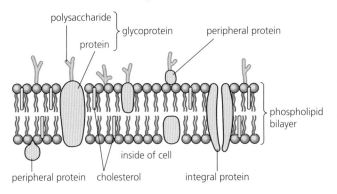

4 Compare Figure 2.8, page 49 and Figure 2.9, page 50. The difference is the permeability of the membrane traversed. In facilitated diffusion, the permeability changes in response to the presence of the substance that passes – it triggers the opening of pores through which diffusion occurs.

5 See the enlarged inset in Figure 2.10, page 50.

6

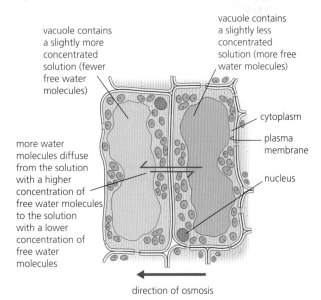

direction of osmosis

7 Uptake of ions is by active transport involving metabolic energy (ATP) and protein 'pump' molecules located in the plasma membranes of cells. As with all aspects of metabolism, this is a temperature-sensitive process and occurs more speedily at 25 °C than at 5 °C.

Individual ions are 'pumped' across by specific, dedicated protein molecules. Because there are many more sodium ion pumps than chloride ion pumps, more sodium ions are absorbed.

8 a diffusion
  b osmosis
  c active uptake
  d phagocytosis

9 Lipids may be dissolved in organic solvents, so we can anticipate the partial destruction of a plasma membrane in contact with ethanol.

10

Characteristics of an efficient respiratory surface	Reasons why each influences diffusion
a large, thin surface area	the greater the surface area and the shorter the distance that gases (O_2 and CO_2) have to diffuse, the quicker gas exchange occurs
a ventilation mechanism that moves air (or water) over the respiratory surface	the higher the concentration of oxygen on the 'supply side' of the respiratory membrane, the steeper the concentration gradient, and therefore the quicker gas can diffuse across the surface
a blood circulation that speeds up the removal of dissolved oxygen, with a respiratory pigment that increases the gas-carrying capacity of the blood	the quicker that oxygen is picked up and transported away from the gas exchange surface, the steeper the concentration gradient maintained, and the greater the rate of diffusion

The relationship of these factors is summarised by Fick's Law of Diffusion:

$$\text{rate of diffusion} \div \text{surface area} \times \frac{\text{difference in concentration}}{\text{length of diffusion pathway}}$$

11 Carbon dioxide is an acidic gas which, if it were to accumulate in the blood, would alter the pH of the plasma solution. The normal pH of the blood is 7.4, but for life to be maintained it cannot be allowed to vary more than within the range pH 7.0–7.8. This is largely because blood pH affects the balance of essential ions which are transported in the plasma solution. Efficient removal of respiratory CO_2 at the lungs is as important to life as efficient uptake of O_2.

12 Enzymes work by binding to their substrate molecule at a specially formed pocket on the enzyme – the active site. Most enzymes are large molecules, and the active site takes up a relatively small part of the total volume of the enzyme molecule. Nevertheless, the active site is a function of the overall shape of the globular protein, and if the shape of an enzyme changes for whatever reason, the catalytic properties may be lost.

13 a Working with an excess of substrate molecules (a relatively high concentration of substrate), further increase in the concentration of the substrate has no effect on the rate of reaction – see Figure 2.29, page 69.

b The effect of an increase in the concentration of the enzyme is to increase the rate of reaction. This is because, at any moment, proportionally more substrate molecules are in contact with an enzyme molecule.

14 After three generations, 75% of the DNA would be 'light'.

the experimental results that would be expected if the Meselson–Stahl experiment were carried on for three generations

15 a The sequence of amino acids for which this mRNA sequence

Codon	GGU	AAU	CCU	UUU	GUU	ACU	CAU	UGU
Amino acid	Gly	Asn	Pro	Phe	Val	Thr	His	Cys

codes is found using the genetic code (Figure 2.37, page 000):

b The sequence of bases in the coding strand of DNA from which this mRNA was transcribed would be:

CCA TTA GGA AAA CAA TGA GTA ACA

16

Step	RNA	Role
transcription	messenger RNA (mRNA)	carries copy of the code from a single protein-coding gene from the DNA of a chromosome in the nucleus out into the cytoplasm
translation	messenger RNA (mRNA)	delivers a copy of the code from a single protein-coding gene to a ribosome
	transfer RNA (tRNA)	carries specific amino acids to the site of assembly of a protein, in a ribosome
	ribosomal RNA	ribosome is site of protein synthesis

17 a See *Chromosomes occur in pairs*, page 82.

b Every time the nucleus divides by mitosis in growth and development the daughter cells formed have identical numbers of chromosomes to those of the parent cell from which they originated. This is essential because otherwise different parts of our body might start working to different blueprints. The result would be chaos!

18 The layout of your monohybrid cross will be as in Figure 2.47 (page 87). However, the parental generation (P) will have phenotypes *homozygous round* and *homozygous wrinkled*, and genotypes **RR** and **rr**, respectively.

The gametes the parental generation produce will be **R** and **r**.

The offspring (F_1) will have genotype **Rr,** and be described as *heterozygous round*.

When this generation are crossed (selfed), the gametes of will be ½ **R** + ½ **r**.

From an appropriate Punnett grid, the offspring (F_2) to be expected and the proportions are:

Genotypes	RR	Rr	rr
Ratio	1	2	1
Phenotypes	round	round	wrinkled

3 The voice of the genome

1 *Magnification* is the number of times larger an image is than the specimen. The magnification obtained with a compound microscope depends on which of the lenses you use. For example, using a × 10 eyepiece and a × 10 objective lens (medium power), the image is magnified × 100 (10 × 10). When you switch to the × 40 objective (high power) with the same eyepiece lens, then the magnification becomes × 400. These are the most likely orders of magnification used in your laboratory work.

The resolving power (*resolution*) of the microscope is its ability to separate small objects which are very close together. If two separate objects cannot be resolved they will be seen as one object. Merely enlarging them will not separate them.

Resolution is determined by the wavelength of light. Light is composed of relatively long wavelengths, whereas shorter wavelengths give better resolution. For the light microscope, the resolving power is about 0.2 m. This means two objects less than 0.2 m apart will be seen as one object.

2 A *lipid bilayer* is demonstrated in Figure 2.1 (page 000). In the presence of sufficient lipid, the molecules of lipid arrange themselves as a bilayer, with the hydrocarbon tails facing together. This is the situation in the plasma membrane.

Several organelles of eukaryotic cells have a *double membrane*, including chloroplasts and mitochondria, in which there is an outer and an inner membrane, both of which consist of lipid bilayers.

So the difference between a lipid bilayer and a double membrane lies in the number of lipid bilayers present.

2 A *lipid bilayer* is demonstrated in Figure 2.1 (page 45). In the presence of sufficient lipid, the molecules of lipid arrange themselves as a bilayer, with the hydrocarbon tails facing together. This is the situation in the plasma membrane.

Several organelles of eukaryotic cells have a *double membrane*, including chloroplasts and mitochondria, in which there is an outer and an inner membrane, both of which consist of lipid bilayers.

So the difference between a lipid bilayer and a double membrane lies in the number of lipid bilayers present.

3

	Name	Role
A	nucleus	controls and directs the activities of the cell
B	ribosomes	site of synthesis of proteins that remain in the cell
C	mitochondria	site of part of the reactions of aerobic respiration of sugars
D	lysosomes	contain hydrolytic enzymes that process materials taken in by phagocytosis and of damaged or redundant cellular components
E	Golgi apparatus	plays a part in 'packaging' and directing secretory substances to their correct destination
F	vesicles	bud off from the margins of the Golgi apparatus, and contain chemicals required by the cell
G	RER	site of the packaging of proteins for export from the cell

4 See Figure 3.9, page 104.
5 The electron microscope has powers of magnification and resolution that are greater than those of an optical microscope. The wavelength of visible light is about 500 nm, whereas that of a beam of electrons used in the EM is 0.005 nm. At best, the light microscope can distinguish two points which are 200 nm (0.2 m) apart. The effective resolution of the transmission electron microscope in the study of biological materials is considerably less than the ideal, but under optimum conditions, a resolution of up to 0.10 nm can be obtained – about 2000 times better resolution than the best resolution of light microscopes. Given the sizes of cells, of the organelles they contain and of the membranes that form many of these organelles, it requires the magnification and resolution achieved in transmission electron microscopy to observe cell ultrastructure – in suitably prepared specimens.

6

	Examples from	
	a flowering plant	**b mammal**
organ	stem, leaf, root	heart, liver, lungs,
tissue	ground tissue (parenchyma) xylem (water conducting tissue)	cardiac muscle, blood
undifferentiated cell	cells of the terminal and lateral growing points (called meristematic cells)	embryonic stem cells of early embryo, adult stem cells of most organs (page 123)

7 Nuclear membrane and its pores, nucleolus, chromatin.
8 In metaphase, the nuclear membrane has broken down and the spindle has formed. The chromosomes (each made up of two chromatids) attach by their centromeres to spindle fibres at the equator of the spindle. This is in preparation for being separated (pulled apart) and dragged to the poles of the spindle. Separation and transport processes are aided by the fact that the (typically) 50 000 m length of DNA of a chromosome has been 'supercoiled' to a length of only about 5 m. After all, most nuclei contain many

chromosomes in very close proximity which have to be separated simultaneously without mechanical damage to their DNA.
9 A major event of interphase is the replication of the chromosomes (Figure 3.14, page 109). By the omission of this stage between meiosis I and II, the chromosome numbers are halved in the four cells produced following meiotic cell division.
10 See Figure 3.19, page 115.

	Mitosis – a replicative division	**Meiosis – a reductive division**
product	two identical cells each with the diploid number of chromosomes	four, non-identical cells each with the haploid number of chromosomes
consequences or significance	permits growth and repair within multicellular organisms also, asexual reproduction	contributes to genetic variability by: ■ reducing the chromosome number by half, permitting subsequent fertilisation and the combination of genes of two parents ■ permitting the random assortment of maternal and paternal chromosomes ■ recombinations of segments of individual maternal and paternal homologous chromosomes during crossing over

11

	Actual size	**Drawing**	**Magnification**
sperm	6 m long	85 mm long	$\dfrac{(85 \times 1000)}{6} = \times 14\,167$ i.e. **about $\times$ 14 000**
oocyte	135 m diameter	20 mm diameter	$\dfrac{(20 \times 1000)}{135} = \times 148$ i.e. **about $\times$ 150**

12 See *Pollination and fertilisation*, page 121.
13

	a Totipotent cells	**b Pluripotent cells**	**c Multipotent cells**
locations	8-cell blastocyst	cells of the inner cell mass of the early embryo	adult stem cells, a few in most organs
roles	create the early embryo	make up the bulk of the embryo/fetus tissues	replace dead or damaged cells

14 See *Glossary*, on the DL Student website.
15 This is a personal issue which needs handling with great sensitivity. When you discuss a response to this question with your peers, it is best if each individual explains their own approach and is listened to carefully. It may be that a consensus concerning a suitable response can be arrived at – but there are circumstances over issues of this sort where individuals cannot agree. In this situation, to understand the approach of those you do not agree with is highly desirable – rather than conflict!
16 See *Glossary*, on the DL Student website.
17

phenotypes	white	light	medium	dark	black
genotypes	1 aabb	2 Aabb 2 aaBb	1 AAbb 4 AaBb 1 aaBB	2 AaBB 2 AABb	1 AABB
ratio	1	4	6	4	1

18 See *Glossary*, on the DL Student website.
19 Cancer cells divide uncontrollably, forming a tumour. The cells of a malignant tumour invade and damage surrounding organs. Cancer cells may detach and be transported by the blood circulation (metastasise) to distant sites in the body. Normal cells do not behave in this way.

   The switch to uncontrolled growth of cancer cells is caused by the accumulation of harmful genetic changes (mutations) in a range of genetically controlled mechanisms that regulate the cell cycle.

   The cell cycle is a regulatory loop. It ensures that DNA is faithfully copied and that the replicated chromosomes move to the new (daughter) cells. Further, cell division is restricted – only a selected range of cells can divide again. Normally, if any cell is severely damaged, or grows abnormally, or becomes redundant in the further development of an organ, it undergoes programmed cell death. Thus, many controls prevent cancerous cell behaviour in normal cells.

   So cancer cells arise as a result of the coincidence of mutations in several 'normal growth and behaviour' genes. Once some cancers are established, malignant tumour cells typically release a specific protein that dictates the formation of a network of blood vessels supplying the tumour. This occurs at the expense of the supply of nutrients to surrounding healthy cells. This enhances abnormal cell division and growth, and leads to further metastasis.

4 Biodiversity and natural resources

1 a See **A* Extension 3.2: The limitations of the electron microscope** on the DL Student website.
  b For TEMs of thin sections, cells or tissues are killed and chemically 'fixed' in a lifelike condition, embedded in a resin medium and sectioned to less than 2 m thick. The fragile specimens are floated off the knife point on to a supporting fine copper grid. Sections are stained with electron-dense materials, such as solutions of lead salts (for lipid-containing membranes) or uranyl salts (for protein and nucleic acids). The dried grids with the stained sections are transferred to the EM. The alternative approach, involving freeze-etching, is outlined in **A* Extension 3.2: The limitations of the electron microscope** on the DL Student website.

2

	Name	Role
A	nucleus	controls and directs the activity of the cell
B	Golgi apparatus	plays a part in packaging and directing secretory substances to their correct destination
C	mitochondrion	site of some of the reactions of aerobic respiration of sugars
D	RER	site of the synthesis and packaging of proteins for export from the cell
E	permanent vacuole	a fluid filled space within the cytoplasm, possibly where certain metabolites are stored or disposed of
F	position of tonoplast	the selectively permeable membrane around the vacuole – controls entry and exit of metabolites

3 a See *Glossary*, on the DL Student website.
  b Lichens are symbiotic organisms consisting of an alga and a fungus (for example).
4

Starch		Cellulose
	Similarities	
polymer of glucose	**what each is made of**	polymer of glucose
	Differences	
mixture of 2 polymers of α glucose combined together	**components**	polymer of β glucose
amylose, a straight-chain polymer of α 1,4-glycosidic linkages + amylopectin, a branched-chain of α 1,4-glycosidic linkages + branches of α 1,6-glycosidic linkages; both adopt a helical shape	**structure and shape**	single polymer of β 1,4-glucosidic linkages; a straight chain molecule with an excess of hydrogen bonds linking the parallel chains.
relatively easily hydrolysed to glucose	**hydrolysis to glucose**	strong molecule, not easily hydrolysed back to glucose
food reserve	**role**	a structural and support molecule

5 A pit is a minute, wall-free area in the cell wall of a plant cell.
   A plasmodesma is one or more cytoplasmic 'threads' (basically endoplasmic reticulum) passing transversely through plant cell walls, connecting the cytoplasm of adjacent cells.

6

	Similarities	Differences
Position	Occur in the outer parts of the stem.	Xylem restricted to the vascular bundles, on the inner side. Sclerenchyma occurs between xylem vessels, and also as a 'cap' at the outer part of many bundles. Occasionally, fibres are also found below the epidermis, in older stems.
Structure	Walls of cellulose impregnated with lignin. Pits are typically found in their walls.	Xylem vessels have end-walls dissolved; they exist as continuous tubes. Sclerenchyma fibres have pointed, tapering ends that interlock.
Function	Support the stem.	Only xylem vessels are concerned in the conduction of water by means of a continuous column flowing from roots to leaves (known as the transpiration stream).

7 These are ideal environmental conditions favouring the evaporation of water from a moist surface, and for maintaining a diffusion gradient for evaporation. Heat provides the energy for evaporation, dry conditions and the presence of wind maintain the gradient favouring diffusion of water vapour molecules.

8 The harvesting of crops removes mineral elements, and cropping interrupts the natural recycling of nutrients. Loss of nutrients can be offset by the addition of manure or compost (natural organic waste material) or artificial fertilisers (chemicals).

Arable crop yields in the UK have increased greatly during the past 50 years. This is because of the use of improved varieties of wheat, barley, oats and maize. Also, the practice of providing essential mineral ions to the soil of arable farms, by the regular and timely application of fertilisers, has been essential. Minerals lost from the soil during crop growth and harvesting have been entirely restored.

Artificial fertilisers provide minerals in a form that is immediately or quickly made available to the growing plants. The minerals present in the fertiliser granules mostly dissolve speedily and are taken up by the roots from the soil solution. Consequently, these fertilisers are applied at the start of peak growing periods of crop plants, usually after weed plants have been killed off. This minimises the loss of soluble fertilisers to competitor weeds or into the water table below. Modern plant varieties mostly require relatively high levels of ions at key times in their growth in order to provide exceptional yields at the end of the season.

Farmyard manure (particularly from the housing of herds of cows in sheltered yards over winter) may be spread onto arable fields and ploughed in. Saprotrophic organisms in the soil break down the dung and straw, releasing mineral ions. These are taken up by the crop plants as they grow. Decay of manure is a slow process, and ions are released over an extended period, not necessarily when most required by the crop plant. However, the organic matter normally leads to a build-up of the soil structure, improving aeration and drainage. Soil rich in organic matter is darker in colour and absorbs more heat; yet it retains water well. However, manure may increase the weed seeds present in the soil.

9 For details of fibre (sclerencyhma) structure, see Figure 4.7 (page 148).

Surface hair cells are parenchyma cells with walls of cellulose only.

10 A *renewable resource* is one that can be replenished by natural processes at a rate comparable to the rate at which it is exploited or consumed. The list of renewable resources is long (for example, solar power, wind power, hydropower, wood and fibre products, and so on). However, it does not include fossil fuels (these are non-renewable resources, such as oil and coal) because these were laid down under exceptional geological conditions a very long time ago – conditions which may not be repeated in the realistic future.

11 Is it possible that the natural insecticide molecules will leave residues in plant matter that may enter the human food chain in vegetables that have been genetically engineered? Will any of these molecules be harmful to humans? (Bear in mind that many plants adopt 'chemical warfare' against browsers on a significant scale, and we may be consuming some of these, currently, with little or no ill-effects.)

12 Precisely defined and internationally agreed, they facilitate co-operation between observers by ensuring the exact species that is being investigated and reported on.

13 The Simpson Diversity Index for this habitat:

Species	n (number of individuals)	$n - 1$	$n(n - 1)$
groundsel	45	44	1980
shepherd's purse	40	39	1560
dandelion	10	9	90
total (N)	95		

$n(n - 1) = 1980 + 1560 + 90 = 3630$

$$D = 95 \times \frac{94}{3630}$$

$$= \frac{8930}{3630}$$

$$= \mathbf{2.46}$$

14

15 Tourism is a major source of income for Sabah, and tours and visits arranged with responsible organisations safeguard the local environment from harm, and help to support conservation initiatives.

16 Charles Darwin, together with virtually everyone else working in science at the time the 'Origin of Species' was published, knew nothing of Mendel's discovery of the principles of modern genetics (page 84). Chromosomes had not been reported, and the existence of genes, alleles and DNA were unknown.

Instead, biologists believed in 'blending inheritance' to account for the similarities and differences between parents and offspring. According to this explanation, an offspring was a 'blend' or average of the characteristics of the parents. The outcome of the blending of the characteristics of parents in their offspring, if it happened, would be increasing uniformity. That is, genetic variation (which is essential for natural selection) would actually be reduced. Today, modern genetics has shown us that blending generally does not occur, and that there are several ways by which genetic variations arise in gamete formation and fertilisation.

17 a Exposure of pathogenic bacteria to sub-lethal doses of antibiotic may increase the chances of resistance developing in that population of pathogens.

b By varying the antibiotics used, there is increased likelihood of killing all the pathogens in a population, including any now resistant to the previous antibiotics used. This approach works until multiple-resistance strains have evolved, such as in strains of *Clostridium difficile* and *Staphylococcus aureus*, for example.

18 Viruses are not living, as biologists know the term; they are not cellular in organisation. Far too small to be seen by light microscopy (typically 20–300 nm in size), we only know about their structure because of the electron microscope. A virus consists of a strand of nucleic acid (either DNA or RNA), wrapped up in a protein coat. A virus can only reproduce by getting inside a host cell, there taking over the enzymic machinery and switching it to the exclusive production of virus particles. Outside the host cell viruses are described as 'crystalline'.

The fungi are a kingdom of eukaryotic organisms, and include the moulds, yeasts, mildews, mushrooms, puffballs and rusts. The fungus 'body' is known as a mycelium, and consists of fine, tubular hyphae. These, though plant-like, have walls not of cellulose but of a polysaccharide called chitin. The nutrition of fungi is heterotrophic; many feed on dead organic matter (saprotrophic nutrition) but others are parasites, particularly of plants. Fungi reproduce by the production of spores which are typically dispersed by air currents.

19 Biochemical analyses of RNAs extracted from cells.

20 Profligate consumption of the world's finite resources, pollution of much of the environment, destruction of habitats, and the enhancement of global warming.

21 You may have the opportunity to talk with a voluntary or professional Warden service, who might also be prepared to provide a guided walk or lecture presentation about current management policy and the potential of the Reserve that is being conserved.

Index

The Publishers would like to thank the following for permission to reproduce copyright material:

Photo credits
p. 7 l, Andrew Lambert Photography/Science Photo Library; **p. 7 r**, Andrew Lambert Photography/Science Photo Library; **p. 8**, Biophoto Associates; **p. 11**, Andrew Lambert Photography/Science Photo Library; **p. 14**, © Gene Cox; **p. 17**, CNRI/Science Photo Library; **p. 24**, BSIP VEM/Science Photo Library; **p. 35**, Stéfan Rousseau/PA Photos; **p. 47 r**, Don W. Fawcett/Science Photo Library; **p. 47 l**, NIBSC/Science Photo Library; **p. 60**, © Gene Cox; **p. 82 l**, Biophoto Associates; **p. 82 r**, Biophoto Associates; **p. 84**, Nigel Cattlin/FLPA; **p. 89**, Dr P. Marazzi/Science Photo Library; **p. 99**, J. C. Revy/Science Photo Library; **p. 100 l**, Eye of Science/Science Photo Library; **p. 100 r**, Prof. P. Motta/Dept. of Anatomy/University 'La Sapienza', Rome/Science Photo Library; **p. 102 t**, Kevin Mackenzie, School of Medical Science, Aberdeen University; **p. 102 b**, K. R. Porter/Science Photo Library; **p. 103 t**, Biophoto Associates; **p. 103 b**, Biophoto Associates; **p. 104**, Biophoto Associates; **p. 105**, John Greenhorn, School of Medicine, Aberdeen University; **p. 107**, © Gene Cox; **p. 110, all**, Biophoto Associates; **p. 112**, Biophoto Associates; **p. 114**, © Gene Cox; **p. 122**, Dr Keith Wheeler/Science Photo Library; **p. 124**, Dr G. Moscoso/Science Photo Library; **p. 136 br**, Jim Strawser/Alamy; **p. 136 bl**, Juniors Bildarchiv/Alamy; **p. 136 tr**, William Ervin/Science Photo Library; **p. 136 tl**, Hubert M.-L./Still Pictures; **p. 143**, Biophoto Associates; **p. 145**, Dr Kari Lounatmaa/Science Photo Library; **p. 146**, Biophoto Associates/Science Photo Library; **p. 150**, © Gene Cox; **p. 151 l**, Biophoto Associates; **p. 151 r**, Biophoto Associates; **p. 152**, © Gene Cox; **p. 153**, © Gene Cox; **p. 158**, Dr Richard Johnson, formerly of Aberdeen University; **p. 166**, Dr C. J. Clegg; **p. 169**, Jim Zipp/Science Photo Library; **p. 177**, © Martin Harvey/NHPA; **p. 188**, The Natural History Museum/Alamy; **p. 194**, Robert Bird/Alamy; **p. 201 b**, Jack Sullivan/Alamy; **p. 201 tr**, NRT-Helena/Alamy; **p. 201 br**, Thomas Schneider/Agstockusa/Science Photo Library.
l = left, r = right, t = top, b = bottom, m = middle

Acknowledgements
The Publishers wish to thank the following for permission to reprint copyright material in this book, as listed below:

Figure 1.07: Lubert Stryer, Glycogen (adapted), from *Biochemistry*, 3rd edition (W. H. Freeman, 1988); **Figure 1.20**: J. H. Green, Changing blood pressure in the circulation system (adapted), from *An Introduction to Human Physiology* (Oxford University Press, 1976); **Figure 1.27**: Premature death by causes in the UK, 2004, from British Heart Foundation statistics database, www.heartstats.org, reproduced by permission of BHF Health Promotion Research Group; **Figure 1.29:** Figure (adapted) from *Salters-Nuffield Advanced Biology AS: Student Book* (Heinemann Educational Publishers, 2005), reproduced by permission of Pearson Education; **Figure 1.30**: Gerard Tortora and Sandra Grabowski, Figure 20.19 (adapted), from *Principles of Anatomy and Physiology*, 9th edition (John Wiley & Sons, 1999); **Figure 1.31**: Brian Fox and Allan G. Cameron, Figure 3.1 from *Food Science, Nutrition and Health*, 7/Rev.Ed (Hodder Arnold, 2006), reproduced by permission of Edward Arnold (Publishers) Ltd; **Figure 1.32**: World Health Organization, Correlation studies in the epidemiology of CHD, data from *Report of the WHO Study Group, WHO Technical Report Series, No 797, Geneva*, in Andrew Edmonson and David Druce, *Advanced Biology – Statistics* (Oxford University Press, 1996); **Figure 1.34**: Richard Ford, David Rose and Patrick Foster, 'Crackdown on middle class wine drinkers', from *The Times* (5 June 2007), reproduced by permission of News International Syndication; **Figure 1.35**: Helen Carter, 'Girls gripped by worry over looks', from *The Guardian* (16 November 1998), copyright Guardian News & Media Ltd 1998, reproduced by permission of the publisher; **End-of-topic test, Chapter 1, Question 5**: J. H. Green, Graph – how blood pressure changes throughout the circulation (adapted), from *An Introduction to Human Physiology* (Oxford University Press, 1976); **Figure 2.14**: C. J. Clegg and D. G. Mackean, from *Advanced Biology: Principles and Application*, second edition (John Murray, 2000), reproduced by permission of John Murray (Publishers) Ltd; **Figure 2.37**: E. J. Wood, C. A. Smith and W. R. Pickering, The genetic code – a universal code (adapted), from *Life Chemistry and Molecular Biology* (Portland Press, 1997), © Portland Press Ltd; **Figure 2.43**: Office for National Statistics, Graph (adapted) from 'The Heights and Weights of Adults in Great Britain', © Crown copyright 1999, reproduced by permission; **Figure 2.51**: C. J. Clegg, Figure 3.29 (adapted) from *Genetics and Evolution* (John Murray, 1999), reproduced by permission of John Murray (Publishers) Ltd; **End-of-topic test, Chapter 2, Question 3**: Dianne Gull, 'Enzymes fast and flexible', from *Biological Science Review*, 19.1 (Philip Allan Updates, 2006); **End-of-topic test, Chapter 2, Question 15 (DL disk only)**: Lynn Burnet, Pedigree Chart, from *Exercises in Applied Genetics* (Cambridge University Press, 2000), reproduced by permission of the publisher; **Figure 3.13**: J. Z. Young, Diagram, TS small intestine (adapted) from *Life of Mammals* (Oxford University Press, 1957), reproduced by permission of the publisher; **Figure 3.25**: Jack Cohen, The stages of fertilisation (adapted), from *Reproduction* (Butterworth, 1977); **End-of-topic test, Chapter 3, Question 13**: Mark Henderson, 'Breakthrough as stem cells are produced from skin not embryos' and 'A valuable development, but a moral case is exaggerated', from *The Times* (21 November 2007), reproduced by permission of News International Syndication; **Figure 4.05**: J. Bonner and A. W. Galston, The chemistry of cellulose (adapted), from *Principles of Plant Physiology* (W. H. Freeman, 1953); **Figure 4.07**: A. W. Robards, Figure 12.9 and Figure 12.11 (adapted) from *Electron Microscopy and Plant Ultrastructure* (McGraw-Hill Education, 1970), reproduced by permission of the author; **Figure 4.08**: C. J. Clegg, Figure 1.17(ii) (adapted) from *Anatomy and Activity of Plants* (John Murray, 1989), reproduced by permission of John Murray (Publishers) Ltd; **Figure 4.16**: Steven Vogel, Figure 15.1(a), Figure 15.2(a), from *Comparative Biomechanics* (Princeton University Press, 2003), reproduced by permission of the publisher; **Figure 4.22**: C. J. Clegg, D. G. Mackean et.al., Figure 16.3 (adapted) from *Advanced Biology: Principles and Applications: Study Guide* (Hodder Murray, 1996), reproduced by permission of John Murray (Publishers) Ltd; **Figure 4.30**: Stuart L. Pimm and Clinton Jenkins, Maps (adapted) from 'Sustaining the Variety of Life', from *Scientific American*, 293:3 (2005), © Scientific American Inc. All rights reserved; **Figure 4.31**: Marc Ancrenaz et al., Map of Sabah, from 'Aerial Surveys Give New Estimates for Orangutans in Sabeh, Malaysia', *PLoS Biology* 3(1), 2005, © 2004 M. Ancrenaz, O. Gimenez, L. Ambu, K. Ancrenaz, P. Andau et al; **Figure 4.35:** J. Z. Young, External ear size of hares in relation to latitude (figure, adapted), from *Life of Mammals* (Oxford University Press, 1990); **Figure 4.41**: Alastair Gray (Editor), 'The changing pattern of the estimated world human population', from *World Health and Disease* (Open University Press, 1985), reproduced by permission of the publisher; **Figure 4.42**: Nick Meo, 'Rainforest Eden faces death by chainsaw – to make margarine' (adapted), from *The Times* (2 February 2006); **End-of-topic test, Chapter 4, Question 3**: H. Opik, S. Rolfe and A. Willis, data from *The Physiology of Flowering Plants* (Cambridge University Press, 2005); **End-of-topic test, Chapter 4, Question 10**: C. J. Clegg, Report on biological sources of plastics (adapted) from *Microbes in Action* (Hodder Murray, 2002), reproduced by permission of John Murray (Publishers) Ltd.

Every effort has been made to establish copyright and contact copyright holders prior to publication. If contacted, the publisher will be pleased to rectify any omissions or errors at the earliest opportunity.